In the social sciences norms are sometimes taken to play a key explanatory role. Yet norms differ from group to group, from society to society, and from species to species. How are norms formed and how do they change?

This "state-of-the-art" collection of essays presents some of the best contemporary research into the dynamical processes underlying the formation, maintenance, metamorphosis, and dissolution of norms. The volume combines formal modeling with more traditional analysis, and considers biological and cultural evolution, individual learning, and rational deliberation.

T0291502

The dynamics of norms

Cambridge Studies in Probability, Induction, and Decision Theory

General editor: Brian Skyrms

Advisory editors: Ernest W. Adams, Ken Binmore, Jeremy Butterfield, Persi Diaconis, William L. Harper, John Harsanyi, Richard C. Jeffrey, Wolfgang Spohn, Patrick Suppes, Amos Tversky, Sandy Zabell

The dynamics of norms

Edited by

Cristina Bicchieri
Carnegie Mellon University

Richard Jeffrey
Princeton University

Brian Skyrms
University of California, Irvine

CAMBRIDGE
UNIVERSITY PRESS

CAMBRIDGE UNIVERSITY PRESS
Cambridge, New York, Melbourne, Madrid, Cape Town, Singapore, São Paulo, Delhi

Cambridge University Press
The Edinburgh Building, Cambridge CB2 8RU, UK

Published in the United States of America by Cambridge University Press, New York

www.cambridge.org
Information on this title: www.cambridge.org/9780521108744

First published 1997
This digitally printed version 2009

A catalogue record for this publication is available from the British Library

Library of Congress Cataloguing in Publication data

The dynamics of norms/edited by Cristina Bicchieri, Richard Jeffrey,
Brian Skyrms.

p. cm. – (Cambridge studies in probability, induction, and
decision theory)

ISBN 0-521-56062-4

1. Social norms. 2. Social norms – Mathematical models.
I. Bicchieri, Cristina. II. Jeffrey, Richard C. III. Skyrms, Brian.
IV. Series.
GN493.3 D96 1996
306 – dc20   96-13489
                                                        CIP

ISBN 978-0-521-56062-7 hardback
ISBN 978-0-521-10874-4 paperback

Contents

	Preface	*vii*
	Contributors	*ix*
1	**The evolution of strategies in the iterated prisoner's dilemma** Robert Axelrod	1
2	**Learning to cooperate** Cristina Bicchieri	17
3	**On the dynamics of social norms** Pier Luigi Sacco	47
4	**Learning and efficiency in common interest signaling games** David Canning	67
5	**Learning on a torus** Luca Anderlini and Antonella Ianni	87
6	**Evolutive vs. naive Bayesian learning** Immanuel M. Bomze and Jürgen Eichberger	109
7	**Learning and mixed-strategy equilibria in evolutionary games** Vincent P. Crawford	131
8	**Bayesian learning in games: A non-Bayesian perspective** J. S. Jordan	149
9	**Savage–Bayesian agents play a repeated game** Yaw Nyarko	175
10	**Chaos and the explanatory significance of equilibrium:** **Strange attractors in evolutionary game dynamics** Brian Skyrms	199

Preface

In the social sciences *norms* sometimes are taken to play a key explanatory role. But existing norms in turn require their own explanations. Norms differ from group to group, from society to society, and from species to species. How are norms formed and how do they change? These questions lead to a study of the dynamical processes responsible for the establishment, maintenance, meta-morphosis, and dissolution of norms. Biological evolution, cultural evolution, individual learning, and rational deliberation all come into play on the appropriate time scales. This volume presents some of the best contemporary research in this area.

Robert Axelrod's two prisoner's dilemma computer tournaments have focused attention on Anatol Rapoport's winning entry *Tit-for-Tat*, and Axelrod has provided an analysis of the reasons for its success. Here Axelrod describes a computer experiment in which he uses John Holland's genetic algorithm to evolve alternative strategies in essentially the environment of his second tournament. Evolution produces strategies that do better than Tit-for-Tat in this environment. The reader will want to consider these strategies and the significance of their success.

Cristina Bicchieri considers how norms of cooperation can be established in the case of *n*-person social dilemmas that generalize the prisoner's dilemma. She proposes an analysis of social norms as a certain kind of equilibria. She then discusses how cooperation can emerge through individual learning and spread through cultural evolution in a population of bounded rational players.

Pier Luigi Sacco also studies the repeated prisoner's dilemma, using the replicator dynamics of evolutionary game theory to model the evolution of norms within a society. To this he adds a dynamics of constitutional metanorms driven by intersociety comparisons of efficiency.

David Canning investigates quite a different kind of game. Here players have a common interest in implementing successful signaling. Canning studies the evolution of language in the context of repeated signaling games, and establishes conditions under which an efficient signaling system will emerge.

Luca Anderlini and Antonella Ianni introduce location as a modeling factor. They analyze the dynamics of a coordination game played with neighbors, where the players are located on a torus. They report experimental results with cellular automaton hardware and establish long-run convergence to absorbing states analytically.

Immanuel Bomze and Jürgen Eichberger develop the similarities between the replicator dynamics of evolutionary game theory and a kind of naive Bayesian learning process.

Vincent Crawford discusses the interaction of the processes of individual learning and evolution. Even if both processes are modeled by the replicator dynamics, he shows that individual learning generically destabilizes mixed strategy equilibria that would be stable under evolution operating without individual learning.

J. S. Jordan starts with players who do not know each other's payoffs. He describes a "sophisticated Bayesian learning process" that leads players' expectations (but not necessarily their actual play) to converge to a Nash equilibrium. This learning process is recommended to non-Bayesians not as a realistic model of learning but rather as a "powerful heuristic engine."

Yaw Nyarko investigates infinite repeated games in which the players are "Savage–Bayesian" agents. He surveys different restrictions that may be placed on the priors of these players and the convergence results to which these restrictions lead.

Brian Skyrms calls attention to wild nonconvergent behavior possible in the replicator dynamics. In simple, four-strategy evolutionary games, the replicator dynamics can lead to chaos and "strange attractors." He argues that from a dynamical standpoint it is the concept of an attractor rather than that of an equilibrium that is of prime interest.

The variety of dynamical approaches, and the results – both positive and negative – that are reported here should be of interest both to social scientists and to social and political philosophers.

Cristina Bicchieri
Richard Jeffrey
Brian Skyrms

Contributors

Luca Anderlini   St. John's College, Cambridge

Robert Axelrod   School of Public Policy, University of Michigan

Cristina Bicchieri   Department of Philosophy, Carnegie Mellon University

Immanuel M. Bomze   Department of Statistics and Computer Science, University of Vienna

David Canning   Department of Economics, Columbia University

Vincent P. Crawford   Department of Economics, University of California, San Diego

Jürgen Eichberger   Department of Economics, University of Melbourne

Antonella Ianni   Department of Economics, University College London

J. S. Jordan   Department of Economics, University of Minnesota

Yaw Nyarko   Department of Economics, New York University

Pier Luigi Sacco   Department of Economics, University of Florence

Brian Skyrms   Department of Philosophy, University of California, Irvine

1

The evolution of strategies in the iterated prisoner's dilemma

ROBERT AXELROD

Abstract

This paper employs ideas from genetics to study the evolution of strategies in games. In complex environments, individuals are not fully able to analyze the situation and calculate their optimal strategy. Instead they can be expected to adapt their strategy over time based upon what has been effective and what has not. The genetic algorithm is demonstrated in the context of a rich social setting, the environment formed by the strategies submitted to a prisoner's dilemma computer tournament. The results of the evolutionary process show that the genetic algorithm has a remarkable ability to evolve sophisticated and effective strategies in a complex environment.

In complex environments, individuals are not fully able to analyze the situation and calculate their optimal strategy.[1] Instead they can be expected to adapt their strategy over time based upon what has been effective and what has not. One useful analogy to the adaptation process is biological evolution. In evolution, strategies that have been relatively effective in a population become more widespread, and strategies that have been less effective become less common in the population.

Biological evolution has been highly successful at discovering complex and effective methods of adapting to very rich environmental situations. This is accomplished by differential reproduction of the more successful individuals. The evolutionary process also requires that successful characteristics be inherited through a genetic mechanism that allows some chance for new strategies to be discovered. One genetic mechanism allowing new strategies to be discovered is mutation. Another mechanism is crossover, whereby sexual reproduction takes some genetic material from one parent and some from the other.

The mechanisms that have allowed biological evolution to be so good at adaptation have been employed in the field of artificial intelligence. The artificial-intelligence technique is called the "genetic algorithm" (Holland 1975). While

1

other methods of representing strategies in games as finite automata have been used (Rubinstein 1986; Megiddo and Wigderson 1986; Miller 1989; Binmore and Samuelson 1990; Lomborg 1991), the genetic algorithm itself has not previously been used in game-theoretic settings.

This paper will first demonstrate the genetic algorithm in the context of a rich social setting, the environment formed by the strategies submitted to a prisoner's dilemma computer tournament. The results show that the genetic algorithm is surprisingly successful at discovering complex and effective strategies that are well adapted to this complex environment. Next the paper shows how the results of this simulation experiment can be used to illuminate important issues in the evolutionary approach to adaptation, such as the relative advantage of developing new strategies based upon one or two parent strategies, the role of early commitments in the shaping of evolutionary paths, and the extent to which evolutionary processes are optimal or arbitrary.

The simulation method involves the following steps:

1: the specification of an environment in which the evolutionary process can operate;
2: the specification of the genetics, including the way in which information on the simulated chromosome is translated into a strategy for the simulated individual;
3: the design of an experiment to study the effects of alternative realities (such as repeating the experiment under identical conditions to see if random mutations lead to convergent or divergent evolutionary outcomes); and
4: the running of the experiment for a specified number of generations on a computer, and the statistical analysis of the results.

THE SIMULATED ENVIRONMENT

An interesting set of environmental challenges is provided by the fact that many of the benefits sought by living things, such as people, are disproportionately available to cooperating groups. The problem is that while an individual can benefit from mutual cooperation, each one can do even better by exploiting the cooperative efforts of others. Over a period of time, the same individuals may interact again, allowing for complex patterns of strategic interactions (Axelrod and Hamilton 1981).

The prisoner's dilemma is an elegant embodiment of the problem of achieving mutual cooperation and therefore provides the basis for the analysis. In the prisoner's dilemma, two individuals can each either cooperate or defect. The payoff to a player affects its reproductive success. No matter what the other does, the selfish choice of defection yields a higher payoff than cooperation.

	Column Player	
	Cooperate	**Defect**
Cooperate	**3, 3**	**0, 5**
Defect	**5, 0**	**1, 1**

Figure 1. The Prisoner's Dilemma

But if both defect, both do worse than if both had cooperated. Figure 1 shows the payoff matrix of the prisoner's dilemma used in this study.

In many settings, the same two individuals may meet more than once. If an individual can recognize a previous interactant and remember some aspects of the prior outcomes, then the strategic situation becomes an iterated prisoner's dilemma. A strategy would take the form of a decision rule that specified the probability of cooperation or defection as a function of the history of the interaction so far.

To see what type of strategy can thrive in a variegated environment of more or less sophisticated strategies, I conducted a computer tournament for the prisoner's dilemma. The strategies were submitted by game theorists in economics, sociology, political science, and mathematics (Axelrod 1980a). The fourteen entries and a totally random strategy were paired with each other in a round-robin tournament. Some of the strategies were quite intricate. An example is one which on each move models the behavior of the other player as a Markov process, and then uses Bayesian inference to select what seems the best choice for the long run. However, the result of the tournament was that the highest average score was attained by the simplest of all strategies, TIT FOR TAT. This strategy is simply one of cooperating on the first move and then doing whatever the other player did on the preceding move. Thus TIT FOR TAT is a strategy of cooperation based upon reciprocity.

The results of the first round were circulated and entries for a second round were solicited. This time there were sixty-two entries from six countries (Axelrod 1980b). Most of the contestants were computer hobbyists, but there were also professors of evolutionary biology, physics, and computer science, as well as the five disciplines represented in the first round. TIT FOR TAT was again submitted by the winner of the first round, Anatol Rapoport. It won again.

The second round of the computer tournament provides a rich environment in which to test the evolution of behavior. It turns out that just eight of the entries can be used to account for how well a given rule did with the entire set. These eight rules can be thought of as representatives of the full set in the sense that the scores a given rule gets with them can be used to predict the average score the rule gets over the full set. In fact, 98 percent of the variance in the tournament scores is explained by knowing a rule's performance with these eight representatives. So these representative strategies can be used as a complex environment in which to evaluate an evolutionary simulation. What is needed next is a way of representing the genetic material of a population so that the evolutionary process can be studied in detail.

THE GENETIC ALGORITHM

The inspiration for how to conduct simulation experiments of genetics and evolution comes from an artificial-intelligence procedure developed by computer scientist John Holland and called the genetic algorithm (Holland 1975, 1980; Goldberg 1989). For an excellent introduction to the genetic algorithm, see Holland (1992) and Riolo (1992). The idea is based on the way in which a chromosome serves a dual purpose: it provides a representation of what the organism will become, and it also provides the actual material that can be transformed to yield new genetic material for the next generation.

Before going into details, it may help to give a brief overview of how the genetic algorithm works. The first step is to specify a way of representing each allowable strategy as a string of genes on a chromosome that can undergo genetic transformations, such as mutation. Then the initial population is constructed from the allowable set (perhaps by simply picking at random). In each generation, the effectiveness of each individual in the population is determined by running the individual in the current strategic environment. Finally, the relatively successful strategies are used to produce offspring that resemble the parents. Pairs of successful offspring are selected to mate and produce the offspring for the next generation. Each offspring draws part of its genetic material from one parent and part from the other. Moreover, completely new material is occasionally introduced through mutation. After many generations of selection for relatively successful strategies, the result might well be a population that is substantially more successful in the given strategic environment than the original population.

To explain how the genetic algorithm can work in a game context, consider the strategies available for playing the iterated prisoner's dilemma. To be more specific, consider the set of strategies that are deterministic and use the outcomes of the three previous moves to make a choice in the current move. Since there are four possible outcomes for each move, there are $4 \times 4 \times 4 = 64$ different histories of the three previous moves. Therefore, to determine its choice of cooperation or defection, a strategy would only need to determine what to do in each of the situations that could arise. This could be specified by a list of sixty-four C's and D's (C for cooperation and D for defection).

For example, one of these sixty-four genes indicates whether the individual cooperates or defects when in a rut of three mutual defections. Other parts of the chromosome would cover all the other situations that could arise.

To get the strategy started at the beginning of the game, it is also necessary to specify its initial premises about the three hypothetical moves that preceded the start of the game. To do this requires six more genes, making a total of seventy loci on the chromosome.[2] This string of seventy C's and D's would specify what the individual would do in every possible circumstance and would therefore completely define a particular strategy. The string of seventy genes would also serve as the individual's chromosome for use in reproduction and mutation.

There are a huge number of strategies that can be represented in this way. In fact, the number is 2 to the 70th power, which is about 10 to the 21st power.[3] An exhaustive search for good strategies in this huge collection of strategies is clearly out of the question. If a computer had examined these strategies at the rate of one hundred per second since the beginning of the universe, less than 1 percent would have been checked by now.

To find effective strategies in such a huge set, a very powerful technique is needed. This is where Holland's "genetic algorithm" comes in. It was originally inspired by biological genetics, but was adapted as a general problem-solving technique. In the present context, it can be regarded as a model of a "minimal genetics," which can be used to explore theoretical aspects of evolution in rich environments. The outline of the simulation program is given in Table 1. It works in five stages.

1: An initial population is chosen. In the present context the initial individuals can be represented by random strings of seventy C's and D's.
2: Each individual is run in the current environment to determine its effectiveness. In the present context this means that each individual player uses the strategy defined by its chromosome to play an iterated prisoner's dilemma with other strategies, and the individual's score is its average over all the games it plays.[4]
3: The relatively successful individuals are selected to have more offspring. The method used is to give an average individual one mating and to give two matings to an individual who is one standard deviation more effective than the average. An individual who is one standard deviation below the population average would then get no matings.
4: The successful individuals are then randomly paired off to produce two offspring per mating. For convenience, a constant population size is maintained. The strategy of an offspring is determined from the strategies of the two parents. This is done by using two genetic operators: crossover and mutation.

Table 1. The basic simulation

```
* set up initial population with random chromosomes
* for each of 50 generations
  - for each of 20 individuals
    * for each of the 8 representatives
      - use premise part of the chromosome as individual's
        assumption about the three previous moves
      - for each of 151 moves
        * make the individual's choice of cooperate (C) or defect (D)
          based upon the gene that encodes what to do given the three previous moves
        * make the representative's choice of C or D based upon its own
          strategy applied to the history of the game so far
        * update the individual's score based upon the outcome of this
          move (add 3 points if both cooperated, add 5 points if the
          representative cooperated and the individual defected, etc.)
  - reproduce the next generation
    * for each individual assign the likely number of matings
      based upon the scaling function (1 for an average score, 2 for a
      score one standard deviation above average, etc.)
    * for each of 10 matings construct two offspring from the two
      selected parents using crossover and mutation
```

(a) Crossover is a way of constructing the chromosomes of the two offspring from the chromosomes of two parents. It can be illustrated by an example of two parents, one of whom has seventy C's in its chromosome (indicating that it will cooperate in each possible situation that can arise), and the other of whom has seventy D's in its chromosome (indicating that it will always defect). Crossover selects one or more places to break the parents' chromosomes in order to construct two offspring each of whom has some genetic material from both parents. In the example, if a single break occurs after the third gene, then one offspring will have three C's followed by sixty-seven D's, while the other offspring will have three D's followed by sixty-seven D's.

(b) Mutation in the offspring occurs by randomly changing a very small proportion of the C's to D's or vice versa.

5: This gives a new population. This new population will display patterns of behavior that are more like those of the successful individuals of the previous generation and less like those of the unsuccessful ones. With each new generation, the individuals with relatively high scores will be more likely to pass on parts of their strategies, while the relatively unsuccessful individuals will be less likely to have any parts of their strategies passed on.

The computer simulations were done using a population size of twenty individuals per generation. Levels of crossover and mutation were chosen that averaged one crossover and one-half mutation per chromosome per generation. Each game consisted of 151 moves, the average game length used in the tournament. With each of the twenty individuals meeting eight representatives, this made for about twenty-four thousand moves per generation. A run consisted of fifty generations. Forty runs were conducted under identical conditions to allow an assessment of the variability of the results.

The results are quite remarkable: from a strictly random start, the genetic algorithm evolved populations whose median member was just as successful as the best rule in the tournament, TIT FOR TAT. Most of the strategies that evolved in the simulation actually resemble TIT FOR TAT, having many of the properties that make TIT FOR TAT so successful. For example, five behavioral alleles in the chromosomes evolved in the vast majority of the individuals to give them behavioral patterns that were adaptive in this environment and mirrored what TIT FOR TAT would do in similar circumstances. These patterns are:

1: Do not rock the boat: continue to cooperate after three mutual cooperations (which can be abbreviated as C after RRR).
2: Be provocable: defect when the other player defects out of the blue (D after RRS).
3: Accept an apology: continue to cooperate after cooperation has been restored (C after TSR).
4: Forget: cooperate when mutual cooperation has been restored after an exploitation (C after SRR).
5: Accept a rut: defect after three mutual defections (D after PPP).

The evolved rules behave with specific representatives in much the same way as TIT FOR TAT does. They did about as well as TIT FOR TAT did with each of the eight representatives. Just as TIT FOR TAT did, most of the evolved rules did well by achieving almost complete mutual cooperation with seven of the eight representatives. Like TIT FOR TAT, most of the evolved rules do poorly only with one representative, called Adjuster, which adjusts its rate of defection to try to exploit the other player. In all, 95 percent of the time the evolved rules make the same choice as TIT FOR TAT would make in the same situation.

While most of the runs evolve populations whose rules are very similar to TIT FOR TAT, in eleven of the forty runs, the median rule actually does substantially better than TIT FOR TAT.[5] In these eleven runs, the populations evolved strategies that manage to exploit one of the eight representatives at the cost of achieving somewhat less cooperation with two others. But the net effect is a gain in effectiveness.

This is a remarkable achievement because to be able to get this added effectiveness, a rule must be able to do three things. First, it must be able to discriminate between one representative and another based upon only the behavior the other player shows spontaneously or is provoked into showing. Second, it must be able to adjust its own behavior to exploit a representative that is identified as an exploitable player. Third, and perhaps most difficult, it must be able to achieve this discrimination and exploitation without getting into too much trouble with the other representatives. This is something that none of the rules originally submitted to the tournament were able to do.

These very effective rules evolved by breaking the most important device developed in the computer tournament, namely, to be "nice," never to be the first to defect. These highly effective rules always defect on the very first move, and sometimes on the second move as well, and use the choices of the other player to determine what should be done next. The highly effective rules then had responses that allowed them to "apologize" and get to mutual cooperation with most of the unexploitable representatives, and they had different responses that allowed them to exploit a representative that was exploitable.

While these rules are highly effective, it would not be accurate to say that they are better than TIT FOR TAT. While they are better in the particular environment consisting of fixed proportions of the eight representatives of the second round of the computer tournament, they are probably not very robust in other environments. Moreover, in an ecological simulation these rules would be destroying the basis of their own success as the exploited representatives would become a smaller and smaller part of the environment (Axelrod 1984, pp. 49–52 and 203–5). While the genetic algorithm was sometimes able to evolve rules that are more effective than any entry in the tournament, the algorithm was able to do so only by trying many individuals in many generations against a fixed environment. In sum, the genetic algorithm is very good at what actual evolution does so well: developing highly specialized adaptations to specific environmental settings.

In the evolution of these highly effective strategies, the computer simulation employed sexual reproduction, where two parents contributed genetic material to each offspring. To see what would happen with asexual reproduction, forty additional runs were conducted in which only one parent contributed genetic material to each offspring. In these runs, the populations still evolved toward rules that did about as well as TIT FOR TAT in most cases. However, the asexual runs were only half as likely to evolve populations in which the median member was substantially more effective than TIT FOR TAT.[6]

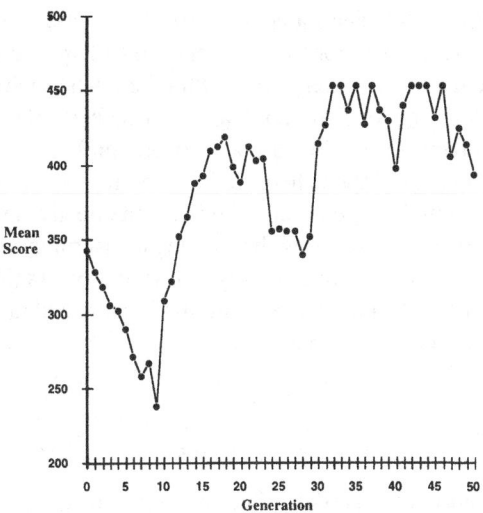

Figure 2. Prisoner's Dilemma Evolving Environment

So far, the simulation experiments have dealt with populations evolving in the context of a constant environment. What would happen if the environment is also changing? To examine this situation, another simulation experiment with sexual reproduction was conducted in which the environment consisted of the evolving population itself. In this experiment each individual plays the iterated prisoner's dilemma with each other member of the population rather than with the eight representatives. At any given time, the environment can be quite complex. For an individual to do well requires that its strategy achieve a high average effectiveness with the nineteen other strategies that are also present in the population. Thus as the more effective rules have more offspring, the environment itself changes. In this case, adaptation must be done in the face of a moving target. Moreover, the selection process is frequency dependent, meaning that the effectiveness of a strategy depends upon what strategies are being used by the other members of the population.

The results of the ten runs conducted in this manner display a very interesting pattern. For a typical run, see Figure 2. From a random start, the population evolves away from whatever cooperation was initially displayed. The less cooperative rules do better than the more cooperative rules, because at first there are few other players who are responsive – and when the other player is unresponsive, the most effective thing for an individual to do is simply defect. This decreased cooperation in turn causes everyone to get lower scores as mutual defection becomes more and more common. However, after about ten

9

or twenty generations, the trend starts to reverse. Some players evolve a pattern of reciprocating what cooperation they find, and these reciprocating players tend to thrive because they can do very well with others who reciprocate without being exploited for very long by those who just defect. The average scores of the population then start to increase as cooperation based upon reciprocity becomes better and better established. So the evolving social environment led to a pattern of decreased cooperation and decreased effectiveness, followed by a complete reversal based upon an evolved ability to discriminate between those who will reciprocate cooperation and those who will not. As the reciprocators do well, they spread in the population, resulting in more and more cooperation and greater and greater effectiveness.

CONCLUSIONS

1: The genetic algorithm is a highly effective method of searching for effective strategies in a huge space of possibilities. Following Quincy Wright (1977, pp. 452–54), the problem for evolution can be conceptualized as a search for relatively high points in a multidimensional field of gene combinations, where height corresponds to fitness. When the field has many local optima, the search becomes quite difficult. When the number of dimensions in the field becomes great, the search is even more difficult. What the computer simulations demonstrate is that the minimal system of the genetic algorithm is a highly efficient method for searching such a complex multidimensional space. The first experiment shows that even with a seventy-dimensional field of genes, quite effective strategies can be found within fifty generations. Sometimes the genetic algorithm found combinations of genes that violate the previously accepted mode of operation (not being the first to defect) to achieve even greater effectiveness than had been thought possible.
2: Sexual reproduction does indeed help the search process. This was demonstrated by the much-increased chance of achieving highly effective populations in the sexual experiment compared to the asexual experiment.[7]
3: Some aspects of evolution are arbitrary. In natural settings, one might observe that a population has little variability in a specific gene. In other words, one of the alleles for that gene has become fixed throughout the population. One might be tempted to assume from this that the allele is more adaptive than any alternative allele. However, this may not be the case. The simulation of evolution allows an exploration of this possibility by allowing repetitions of the same conditions to see just how much variability there is in the outcomes. In fact, the simulations show two reasons why convergence in a population may actually be arbitrary.

(a) Genes that do not have much effect on the fitness of the individual may become fixed in a population because they "hitchhike" on other genes that do (Maynard Smith and Haigh 1974). For example, in the simulations some sequences of three moves may very rarely occur, so what the corresponding genes dictate in these situations may not matter very much. However, if the entire population are descendants of just a few individuals, then these irrelevant genes may be fixed to the values that their ancestors happened to share. Repeated runs of a simulation allow one to notice that some genes become fixed in one population but not in another, or that they become fixed in different ways in different populations.

(b) In some cases, some parts of the chromosome are arbitrary in content, but what is not arbitrary is that they be held constant. By being fixed, other parts of the chromosome can adapt to them. For example, the simulations of the individual chromosomes had six genes devoted to coding for the premises about the three moves that preceded the first move in the game. When the environment was the eight representatives, the populations in different runs of the simulation developed different premises. Within each run, however, the populations were usually very consistent about the premises: the six premise genes had become fixed. Moreover, within each population these genes usually became fixed quite early. It is interesting that different populations evolved quite different premises. What was important for the evolutionary process was to fix the premise about which history is assumed at the start so that the other parts of the chromosome could adapt on the basis of a given premise.

4: There is a trade-off between the gains to be made from flexibility and the gains to be made from commitment and specialization (March 1991). Flexibility might help in the long run, but in an evolutionary system, the individuals also have to survive in the short run if they are to reproduce. This feature of evolution arises at several levels.

(a) As the simulations have shown, the premises became fixed quite early. This meant a commitment to which parts of the chromosome would be consulted in the first few moves, and this in turn meant giving up flexibility as more and more of the chromosome evolved on the basis of what had been fixed. This in turn meant that it would be difficult for a population to switch to a different premise. Flexibility was given up so that the advantages of commitment could be reaped.

(b) There is also a trade-off between short- and long-term gains in the way selection was done in the simulation experiments. In any given generation there would typically be some individuals that did much better than the average and some that did only a little better than the average. In the short run, the way to maximize the expected performance of the next generation would be to have virtually all of the offspring come from the very best individuals in the present generation. But this would imply a rapid reduction in the genetic variability of the population and a consequent slowing of the evolutionary process later on. If the moderately successful were also given a chance to have some offspring, this would help the long-term prospects of the population at the cost of optimizing in the short run. Thus there is an inherent trade-off between exploitation and exploration, that is, between exploiting what already works best and exploring possibilities that might eventually evolve into something even better (Holland 1975, p. 160).

5: Evolutionary commitments can be irreversible. For example, in most of the populations facing the environment of the eight representatives, the individuals evolved strategies that are very similar to TIT FOR TAT. Since TIT FOR TAT had done best in the computer tournament itself, I did not think that it would be possible to do much better with an evolutionary process. But as noted earlier, in about a quarter of the simulation runs with sexual reproduction, the population did indeed evolve substantially better strategies—strategies that were quite different from TIT FOR TAT. These highly effective strategies defected on the very first move, and often on the second move as well, in order to get information to determine whether the other player was the type that could be exploited or not. The more common populations of strategies cooperated from the beginning and employed reciprocity in a manner akin to TIT FOR TAT. While these more common strategies might easily mutate to try a defection at the start of the game, such behavior would be extremely costly unless the individual already had effective ways of using the information that resulted. Moreover, once the population had evolved to be about as effective as TIT FOR TAT, such mutation would have to be quite effective in order to survive long enough to be perfected. Thus, once the population takes a specific route (in this case toward reciprocity), it can easily become trapped in a local maximum. Indeed, only the fact that enough simulation runs were conducted led to the discovery that in this particular environment reciprocity was only a local maximum, and that something better was in fact possible. In a field situation, such a discovery might not be possible, since there might be essentially just one gene pool.

TOPICS AMENABLE TO SIMULATION

The methodology for the genetic simulation developed in this paper can be used to explore learning processes in game-theoretic settings. Here is a list of issues that can be studied with genetic simulations, inspired by analogs to evolutionary biology:

1. Mutation. The simulation approach developed here suggests that there is an inherent trade-off for a gene pool between exploration of possibilities (best done with a high mutation rate) and exploitation of the possibilities already contained in the current gene pool (best done with a low mutation rate). This in turn suggests the advantage of having mutation rates adapt to the rate of change in the environment.[8]

2. Crossover. In sexual reproduction, crossover serves to give each offspring genetic material from both parents. Crossover rates that are too low would frequently give whole chromosomes of genetic material from a single parent to an offspring. But crossover rates that are too high would frequently split up coadapted sets of alleles that are on the same chromosome. Perhaps the existence of a multiplicity of chromosomes (rather than one long chromosome) is more than a mechanical convenience but is an adaptation to the need for low crossover rates without the disadvantage of having each offspring being likely to get genetic material from only one parent.

3. Inversion. Inversion changes the order of the genes in a chromosome. It can bring coadapted sets of alleles closer together on the chromosome so that they will be split apart by crossover less often. How is the ideal rate of inversion determined?

4. Coding principles. Biological chromosomes are known to contain material that does not directly code for proteins but performs other roles, such as marking the boundaries of genes, or perhaps provides no function at all. Genetic material may also appear in highly redundant form in the chromosome. Genetic simulation experiments might shed new light on the theoretical implications of various coding schemes and their possible role in error reduction and regulation. Or they might show how some genetic material can exist as "free riders."

5. Dominant and recessive genes. Mendel's famous experiments demonstrate that dominant and recessive alleles serve to overcome Darwin's concern that blending of parental characteristics would eliminate the variability of a population. Genetic simulation can be used to explore the implications of these and other genetic mechanisms for the maintenance of population variability in the face of selection pressure for local optimality. In particular, it should be possible to explore just which types of phenotypic features are best coded in terms of dominant and recessive genes, and which are best coded in other systems of genetic expression.

13

6. Gradual vs. punctuated evolution. Genetic simulation experiments might also shed light on the contemporary debate about whether evolution proceeds in gradual steps or whether it tends to move in fits and starts. This type of work might require simulations of tens of thousands of generations, but runs of such length are feasible.

7. Population viscosity. Obstacles to random mating may exist due to geographic or other forces tending to favor subdivisions of the population. Some computer modeling has already been done for models of this type (Boorman and Levitt 1980, pp. 78–87; Tanese 1989), revealing clues about the qualitative features of the spread of a social trait based upon frequency-dependent selection.

8. Speciation and ecological niches. When distinct ecological niches exist, a single species tends to differentiate into two or more species to take advantage of the different opportunities offered by the different niches. In learning terms, differentiation into two or more species means that a new strategy is formed from ideas represented in only part of the total population. Genetic simulation can explore this process by helping to specify the conditions under which the advantages of specialization outweigh the disadvantages of narrower mating opportunities and reduced ecological flexibility. The fundamental point is that thinking about genetics as a simulation problem gives a new perspective on the functioning of learning processes.

The genetic simulations provided in this paper are highly abstract systems. The populations are very small, and the number of generations are few. More significantly, the genetic process has only two operators, mutation and crossover. Compared to biological genetics, this is a highly simplified system. Nevertheless, the genetic algorithm displayed a remarkable ability to evolve sophisticated and effective strategies in a complex environment.

NOTES

1. I thank Stephanie Forrest and Reiko Tanese for their help with the computer programming, Michael D. Cohen and John Holland for their helpful suggestions, and the Harry Frank Guggenheim Foundation and the National Science Foundation for their financial support. Previous versions of this paper were presented at a conference called "Evolutionary Theory in Biology and Economics" at the University of Bielfeld, Germany, November 19–21, 1985, and published in Lawrence Davis, ed., *Genetic Algorithms and Simulated Annealing* (London: Pitman; Los Altos, Calif.: Morgan Kaufman, 1987).
2. The six premise genes encode the presumed C or D choices made by the individual and the other player in each of the three moves before the interaction actually begins.

3. Some of these chromosomes give rise to equivalent strategies, since certain genes might code for histories that could not arise given how loci are set. This does not necessarily make the search process any easier, however.
4. The score is actually a weighted average of its scores with the eight representatives, the weights having been chosen to give the best representation of the entire set of strategies in the second round of the tournament.
5. The criterion for being substantially better than TIT FOR TAT is a median score of 450 points, which compares to TIT FOR TAT's weighted score of 428 with these eight representatives.
6. This happened in five of the forty runs with asexual reproduction compared to eleven of the forty runs with sexual reproduction. This difference is significant at the .05 level using the one-tailed chi-squared test.
7. In biology, sexual reproduction comes at the cost of reduced fecundity. Thus if males provide little or no aid to offspring, a high (up to twofold) average extra fitness has to emerge as a property of sexual reproduction if sex is to be stable. The advantage must presumably come from recombination but has been hard to identify in biology. A simulation model has demonstrated that the advantage may well lie in the necessity to recombine defenses to defeat numerous parasites (Hamilton, Axelrod, and Tanese 1990). Unlike biology, in artificial-intelligence applications, the added (computational) cost of sexuality is small.
8. I owe this suggestion to Michael D. Cohen.

References

Axelrod, Robert. 1980a. "Effective Choice in the Prisoner's Dilemma." *Journal of Conflict Resolution* **24**:3–25.

1980b. "More Effective Choice in the Prisoner's Dilemma." *Journal of Conflict Resolution* **24**:379–403.

1984. *The Evolution of Cooperation.* New York: Basic Books.

, and William D. Hamilton. 1981. "The Evolution of Cooperation." *Science* **211**:1390–96.

Binmore, Ken, and Larry Samuelson. 1990. "Evolutionary Stability in Repeated Games Played by Finite Automata." Center for Research on Economic and Social Theory, working paper 90-17, Department of Economics, University of Michigan, Ann Arbor, Mich.

Boorman, Scott A., and Paul R. Levitt. 1980. *The Genetics of Altruism.* New York: Academic Press.

Goldberg, D. E. 1983. *Computer-aided Gas Pipeline Operation Using Genetic Algorithms and Machine Learning.* Ph.D. dissertation, University of Michigan (Civil Engineering).

1989. *Genetic Algorithms in Search, Optimization, and Machine Learning.* Reading, Mass.: Addison-Wesley.

Grefenstette, John J., ed. 1985. *Proceedings of an International Conference on Genetic Algorithms and Their Applications.* Pittsburgh: Robotics Institute of Carnegie-Mellon University.

Hamilton, William D. 1980. "Sex versus Non-Sex versus Parasite." *Oikos* **35**:282–90.

1982. "Heritable True Fitness and Bright Birds: A Role for Parasites." *Science* **218**:384–87.

, Robert Axelrod, and Reiko Tanese. 1990. "Sexual Reproduction as an Adaptation to Resist Parasites." *Proceedings of the National Academy of Sciences USA* **87**:3566–73.

Holland, John H. 1975. *Adaptation in Natural and Artificial Systems.* Ann Arbor: University of Michigan Press.

1980. "Adaptive Algorithms for Discovering and Using General Patterns in Growing Knowledge Bases." *International Journal of Policy Analysis and Information Systems* **4**:245–68.

1992. "Genetic Algorithms." *Scientific American* **267** (July):66–72.

Lomborg, Bjorn. 1991. "An Evolution of Cooperation." Master's thesis, Institute of Political Science, University of Aarhus, Denmark.

March, James G. 1991. "Exploration and Exploitation in Organizational Learning." *Organizational Science,* 2:71–87.

Maynard Smith, J., and J. Haigh. 1974. "The Hitch-hiking Effect of a Favorable Gene." *Genet. Res., Camb.* 23:23–35.

Megiddo, Nimrod, and Avi Wigderson. 1986. "On Play by Means of Computing Machines." IBM Research Division, BJ 4984 (52161), Yorktown Heights, N.Y.

Miller, John. 1989. "The Coevolution of Automata in the Repeated Prisoner's Dilemma." Working paper, 89-003, Sante Fe Institute, Sante Fe, N. Mex.

Riolo, Rick L. 1992. "Survival of the Fittest Bits." *Scientific American* 267 (July):114–16.

Rubinstein, Ariel. 1986. "Finite Automata Play the Repeated Prisoner's Dilemma." *Journal of Economic Theory* 39:83–96.

Tanese, Reiko. 1989. "Distributed Genetic Algorithms for Function Optimization." Ph.D. dissertation, University of Michigan (Computer Science and Engineering).

Wright, Quincy. 1977. *Evolution and the Genetics of Populations. Vol. 4. Experimental Results and Evolutionary Deductions.* Chicago: University of Chicago Press.

2

Learning to cooperate

CRISTINA BICCHIERI

INTRODUCTION

In recent years there has been growing experimental evidence about cooperative behavior even in the absence of egoistic incentives, such as reputation formation opportunities (Andreoni and Miller 1991; Dawes and Thaler 1988). Moreover, it has been observed that once a cooperative pattern of behavior has been established, people tend to expect it to persist (Caporael et al. 1989). Those authors who claim there exists altruistic behavior tend to credit it to the working of norms of cooperation (Dawes 1980). However, remarkably little has been said about the nature of such norms, the circumstances under which they are likely to arise, and the process through which they may spread to entire populations. In the first part of this paper, I spell out the conditions under which norms of cooperation come into being, and the thesis I propose is that they emerge as equilibria of learning dynamics in small-group interactions. The game-theoretic literature on learning is a rich source of models (Crawford 1989; Crawford and Haller 1990; Fudenberg and Kreps 1989; Jordan 1991; Kalai and Lehrer 1990; Milgrom and Roberts 1991), but many of these models lack an explicit account of the dynamics of the learning process, especially in the more realistic case in which players are boundedly rational and do not have "common priors," nor do they assign probability zero to the same set of events. In this paper I do not offer a formal model of learning. Rather, I give the conditions under which boundedly rational players may learn to cooperate and I provide suggestions as to how more realistic learning models might be built.

An important question to ask is what type of behavioral pattern is likely to emerge in small-group interactions among boundedly rational players who try to communicate to each other their (cooperative) intentions in the simplest possible way. A point I want to make is that strategies like tit for tat are likely to emerge as norms of cooperation in virtue of their simplicity (tit for tat requires

An earlier version of this paper was previously published in *Ethics* 100, no. 4 (1990).

17

limited memory on the part of the player who adopts it, and it is easy to "grasp" by an opponent). The size of the group matters, though. Even if learning can easily occur in two-person interactions, it may be impossible in a large population, where all that one observes is aggregate behavior. Yet norms may emerge through learning in a small group and subsequently spread to a large population by some other mechanism.

In the second part of the paper, I suggest that a cognitive explanation for norm-abiding behavior must be coupled with an evolutionary account of the propagation of norms from small groups to populations. Once a norm is established in a given context, people will tend to apply it to contexts that are perceived as relevantly similar to the original one as a "default rule" instead of making complex cost-benefit calculations. However, the social environment in which people act is crucial in determining the survival, spread, or demise of social norms. A norm will be upheld unless and until it becomes clear that it does not do so well as other patterns of behavior in a given environment. An evolutionary approach is useful in modeling the dynamics of norms, because it shows how behavioral patterns change over time in response to their relative success in a changing environment.

NORMS OF COOPERATION

The class of norms I wish to discuss is that of cooperative norms. These norms play an important role in collective action situations, which are closely related to the n-person prisoner's dilemma. In such games, each player has a dominant strategy and rationality dictates choosing it, irrespective of what it expects other players to choose. Specifically, each person can choose whether or not to cooperate and, since the game is noncooperative, there is no external authority to enforce sanctions on the defectors. Defection is thus costless, whereas cooperation is costly. Typical examples include voting, polluting, littering, saving electricity during a hot summer, and doing volunteer work. In all of these cases, the benefits of cooperation depend upon the number of cooperators, although in most cases this number need not be the totality of those concerned. If enough people vote or refrain from consuming electricity, all will benefit from the outcome. But those who did not register or who kept their air conditioner at full power will benefit even more, since they cannot be excluded from enjoying the product of the collective effort of others, while they did not pay any price to start with. In such cases as littering or polluting, where a small number of defectors is enough to do the damage, nearly universal cooperation is needed for the socially desirable outcome to obtain. It is enough that a few people start throwing garbage on a clean beach to induce newcomers to imitate them, since walking to a distant trash can may seem futile when the beach is already spoiled. In each case, cooperation involves the risk of a net loss: If too many people defect, those who cooperate pay the cost and reap no benefit.

In situations that individuals know will not recur, rational self-interested individuals should, therefore, always defect, even if the collective outcome of joint defection is inefficient. It would be better for everybody to cooperate, but since cooperation is a dominated strategy, any agreement to cooperate would fail to be self-enforcing, since each player would have an incentive to cheat on the others. Then what stands in need of explanation is the fact that prisoner's dilemma–like situations often do not result in disastrous outcomes; instead we observe that, overall, people tend to cooperate. There are occasions on which cooperative behavior is dictated by rationality. When the agent is not anonymous, cooperation may be expected even in the absence of external sanctions, since it may be in the individual's interest to form a reputation for being a cooperative type. If one lives in a small community, it may turn out that it is better to return the favors one receives, to pay one's debts, and in general to avoid exploiting one's neighbors, since once one has a reputation for being an untrustworthy person, one may never again receive help and may be cheated by the rest of the group. Being cooperative in this case is a form of "global maximization," in that one is prepared to forgo a gain now for a greater future benefit. Defection should be expected in all those circumstances in which an individual is anonymous, as is the case with large groups, such as the firms in a competitive industry or the shareholders of a company. Defection should also occur in small groups either when it is known that the group will dissolve at a given future date or when some member of the group plans to leave for good. In the latter case, the departing person will have an incentive to cheat. Similarly, if the members of a community believe that their community is coming to an end, the belief, however ungrounded, may be self-fulfilling, in that all sorts of defections will be rationally justified.

When the aforementioned interactions have a well-defined time limit, they can be represented as finitely repeated prisoner's dilemma games. In these games, whenever players have exactly k-level iterated knowledge of the structure of the game and of players' rationality (where k depends on the length of the game), the unique solution to the game is to defect (Bicchieri 1992). When cooperation occurs, it might be due neither to a change in preferences nor to the fact that people commit themselves to nonexploitative behavior. The fact that one's exploitative behavior is likely to be detected and sanctioned by future ostracism is a powerful deterrent, but these interactions have a known time limit. Cooperation becomes less surprising when we relax the informational assumptions that are sufficient for the classical backward induction solution to obtain. For example, cooperation can result when the players have slight doubts about each other's rationality (Kreps et al. 1982). Suppose one player suspects the other to be "irrational" and to play, say, a tit-for-tat strategy with some small probability $\varepsilon > 0$. If the suspected player knows of the other's suspicion, she has an interest in confirming the suspicion by avoiding all moves that will reveal that she is rational. Thus she will not respond to defection with cooperation, nor will she fail to cooperate following a cooperative move of the other player. Playing "as if" she were a tit-for-tat player, she hopes to induce the opponent to respond "kindly." In this case it is possible to cooperate for a long stretch, the total number of noncooperative plays being bounded from above by a number that depends on ε.[1] The same result obtains if each player is rational and knows that the other is rational, but neither knows that the other knows that she is rational. Then nobody is cheated, but everybody has an interest in acting as if she were being cheated.

There are many possible forms of cooperation that players might choose, and tit for tat is just one of them. For example, in a prisoner's dilemma game repeated 50 times, player 1 may decide to cooperate (C) in the first round, and for the next rounds $N = 2, \ldots, T < 50$ to choose C in period N unless player 2 choose to defect (D) in period $N - 1$. For rounds $N > T$, she will always defect, regardless of the other player's choice. Were 2 to play D in period $N - 1$, 1 will respond with D in period N. She may keep playing D until player 2 chooses C, and then play C again. However, she may signal to player 2 her willingness to cooperate by returning to play C immediately after she played D in the previous round. Or they may alternate in playing C and D. In general, since a cooperative pattern is better for both, there will be several cooperative equilibria. With multiple equilibria, it may be impossible to predict which one will in fact be attained, or whether the players will achieve one at all. To predict that a given cooperative equilibrium will obtain, we have to assume that the players make the "right" probability assessment about the type of players (e.g., tit-for-tat or others) they may face. This requirement is at odds with what happens in real-life interactions, since as a rule one has to convince the opponent that one is likely to be a tit-for-tat player (even if one is not), and it cannot generally be assumed that one's opponent understands what such a strategy is, to say nothing of knowing the probability with which that strategy will be played (Selten and Stoecker 1986). By contrast, game theorists take the probability that a player is, say, tit-for-tat as given and assume it to be common knowledge among the players (Kreps et al. 1982).

Typically, the game-theoretic literature explains cooperative solutions to PD-type games assuming one of the following scenarios: (i) the game is infinitely repeated and players have a low enough discount factor (Fudenberg and Maskin 1986), (ii) the game has a finite number of repetitions and the players assess a high probability that there will be another round of play, (iii) the game is finitely repeated and the players know when it ends, but they have doubts regarding their opponents' rationality (Kreps et al. 1982). In all these scenarios, it is assumed that people will always choose the selfish strategy in a social dilemma. The only way to avoid the resulting deficient equilibrium is to embed the dilemma in a larger context involving utilities that make the dominant strategy no longer dominant, thus allowing cooperative behavior to emerge. However, hypotheses about the necessity of incentive changes are testable. For example, a model of "reciprocal altruism," like the one proposed by Kreps et al. (1982), is falsified by experiments showing that cooperation does not decline when the incentive is removed (Andreoni and Miller 1991).

Likewise, evidence of cooperative behavior was obtained in one-shot PD plays with unknown coplayers by Terhune (1968), Pancer (1973), and Rapoport (1988). In these situations reciprocity, reputation, and sanctions must be ruled out as contributing factors, yet in some of the experiments up to 50 percent of the subjects did cooperate. The most interesting experiments have been performed by Dawes, Orbell, van de Kragt, and others over a period of 10 years and have involved more than a thousand subjects (Dawes 1980). These experiments involve contributing to the provision of a public good and are particularly valuable because the experimenters were extremely careful in preventing the possibility of reciprocity, sanctions, and reputation effects (Caporael et al. 1989). Subjects were strangers, they made a single decision, choices were anonymous, and there was no interaction among members of the group. Yet initial cooperation rates in those newly formed groups were extraordinarily high, within the range of 40 percent to 60 percent cooperation. Moreover, almost no contributor believed that his contribution was critical to obtaining the public good with a probability sufficient to justify the cooperative action, and 67 percent of the "givers" believed that so many others would contribute that their own contribution would be redundant (Caporael et al. 1989).

These experiments demonstrate that the behavior of subjects is consistent with the hypothesis that people are not motivated simply by narrow self-interest, and indeed they lend credibility to the view that social norms—in this case, norms of cooperation—act as independent motivating factors. However, unless one takes norms as primitive, it remains to be explained how cooperative norms have emerged and what sort of mechanisms support them. The questions could be rephrased in the following terms: How does a group of individuals playing a given game over and over converge to some stationary equilibrium pattern of play? Once a pattern of behavior has been established, why do people tend to conform to it?

THE NATURE OF NORMS

A long-standing tradition in the social sciences contrasts instrumental rationality and social norms as alternative ways of explaining action. Rational choice theory defines action as the outcome of a practical inference that takes preferences and beliefs as premises. As it is typically given, an explanation in terms of norms depicts a socialized actor whose behavior is not outcome-oriented, since when acting in accordance with a norm one does not engage in a rational calculation, nor does one pay too much attention to the action's consequences. Attempts at bridging the gap have tried to establish either that social norms are rational, in the sense of being efficient means to achieve individual or social welfare (Arrow 1971) or prevent market failure (Coleman 1989), or that it is rational to conform to norms, thus reducing compliance to utility maximization (Akerlof 1976; Kandori 1989).

The first reductionist strategy makes a typical *post hoc, ergo propter hoc* fallacy, since the mere presence of a social norm does not justify inferring that it is there to accomplish some social function. Besides, this view of norms does not account for the fact that many social norms are inefficient, as in the case of discriminatory norms against women and blacks, or are so rigid as to prevent the fine-tuning that would be necessary to accommodate new cases successfully. Even if a norm is a means to achieve a social end, such as cooperation, retribution, or fairness, usually it is not the sole means. Many social norms are underdetermined with respect to the collective objectives they may serve, nor can they be ordered according to a criterion of greater or lesser efficiency in meeting these goals. Such an ordering would be feasible only if it were possible to show that one norm among others is the best means to attain a given social objective. Often, though, the objectives themselves are defined by means of some norm.

Consider as an example norms of revenge; until not long ago, a Sicilian man who "dishonored" another man's daughter or sister had to make amends for the wrong by marrying the woman, or pay for his rashness with his own life. The objective was to restore the family's lost honor, but the social norms dictating the ways in which this could be done were the only means available to identify honor in those circumstances. One may think that some form of monetary compensation would have worked equally well, if not better, in the case in which a marriage was impossible. It would have spared one, perhaps many, lives. But accepting a monetary compensation was no revenge, and since nobody would ever have accepted such an atonement, nobody would have even thought of offering it. Approving of the man who exacts revenge, calling him a "man of honor," does not necessarily involve approval of the norm as rational or efficient. Even if one thinks a norm unjust and useless, it may be difficult not to conform, since violation involves a collective action problem: Nobody wants to be the first to risk social disapproval by breaking the norm openly. That is why people will often break a norm in private but still pay lip service to it in public.

The second reductionist strategy argues that, provided that conformity to a norm attracts approval and transgression disapproval, conforming is the rational thing to do, since nobody willfully attracts discredit and punishment (Rommetveit 1955; Thibaut and Kelley 1959). If others' approval and disapproval act as external sanctions, we again have a cost-benefit argument (Coleman 1989). However, not all social norms involve sanctions, as is indicated by studies of the differences among societies as to the proportion and kind of norms that are subject to organized sanctions (Diamond 1935; Hoebel 1954). Moreover, sanctioning generally works well in small-group, repeated interactions, where the identity of the participants is known and monitoring behavior is relatively easy. Even in such cases, however, one may face the so-called second-order public-good problem. Imposing negative sanctions on transgressors is in everybody's interest, but the individual who observes a transgression faces a dilemma. She will have to decide whether or not to punish the transgressor, where punishing involves costs and where there is no guarantee that other individuals, when faced with the same dilemma, will impose a penalty on the transgressor. In this case, it seems that upholding a norm depends on the previous solution of the "punisher's dilemma." An answer to this problem has been that there exist "metanorms" that tell people to sanction transgressors of lower-level norms (Axelrod 1986). This solution, however, only shifts the problem one level up: Upholding the metanorm itself requires the existence of a higher-level sanctioning system.

Another consideration weakens the credibility of the view that norms are upheld only because of external sanctions. Often there is nobody around to watch what we do, and we still conform to a norm. In this case, fear of external sanctions cannot be a motivating force (Elster 1989). It is generally agreed that all cases of "spontaneous" compliance with norms are the result of internalization (Scott 1971). People who have developed an internal sanctioning system, for example, feel guilt and shame at behaving in a deviant way. It remains to be explained how internalization comes about or, in a rational choice perspective, why it is rational to internalize a norm. Coleman (1989) has argued in favor of reducing internalization to rational choice, in that it is in the interest of a group to get another group to internalize certain norms. In this case, internalization would be the result of socialization (Gouldner 1960). Internalization can, however, be interpreted in a different way. The approach I want to suggest is a cognitive one and is related to my account of how norms may emerge through learning. Once one has learned to behave in a given way, one will tend to persist in the learned behavior unless it becomes evident that, on average, the cost of upholding the behavior significantly outweighs the benefits. The idea that norms may be "sluggish" is in line with results from social psychology that show that, once a norm has emerged in a group, it will tend to persist and guide the behavior of the group member, even if she is facing a new situation and is isolated from the original group (Sherif 1938).

External sanctions play an important role, especially in the genesis of norms. But once a norm is established, there are many other mechanisms that account for conformity. Furthermore, to say that one conforms because of the negative sanctions involved in nonconformity does not distinguish norm-abiding behavior from an obsession, in which one feels an inner constraint to repeat the same action in order to quiet some "bad" thought, or from an entrenched habit that cannot be shed without great unease. Nor does it distinguish norms from hypothetical imperatives enforced by sanctions, such as the rule that prohibits smoking in public areas. In all these cases, avoidance of the sanctions involved in transgression constitutes a decisive reason to conform, independently of what others do.

The line of argument I wish to pursue favors a different kind of reduction. Assuming that norms are rational, or making it unconditionally rational to conform to a norm, takes norms for granted. Asking why social norms persist through time, or why we tend to conform to them, does not shed any light on the norm formation process, since *how* norms emerge is a different story from why they tend to persist. My thesis is that social norms are the outcome of learning in a strategic interaction context; hence they are a function of individual choices and, ultimately, of individual preferences and beliefs. The view that norms are reducible to the preferences and beliefs of those who follow them is not new. David Lewis (1969), Edna Ullmann-Margalit (1977) and, more recently, Peyton Young (1993) have proposed a game-theoretic account of norms and conventions according to which a norm is broadly defined as a Nash equilibrium. The conventional game-theoretic account has serious limits, though. For one, it is a static description of norms as clusters of self-fulfilling expectations; it cannot explain, nor was it meant to explain, how expectations arose or came to be self-fulfilling. The equilibrium account of norms must be supplemented with a story of how interacting agents learn to recognize a behavioral pattern, how they settle upon a stable pattern, and what sort of behavior is more likely to be sustainable as a norm.

Learning a behavioral pattern must not be confused with socialization, the process through which the newcomer comes to accept an established group's norm. Since my subject is the development of new norms in a group, learning cannot be separated from the emergence of a new norm. The size of the group matters, though. Even if learning can easily occur in two-person interactions, it may be impossible in a large population, where all that one observes is aggregate behavior. Norms may emerge through learning in a small group and subsequently spread to a large population by some other mechanism. In the last section of the paper, I propose an evolutionary account of the propagation of norms from small groups to populations. Finally, an analysis of emergence, as opposed to one stressing the functions fulfilled by social norms, may shed light upon the differences between social norms and other types of injunction, such as hypothetical imperatives, moral codes, and legal norms, as well as upon those characteristics that are common to all social norms, however different they might be.

NORMS AS EQUILIBRIA

Two important points stressed by both Lewis and Ullmann-Margalit are that social norms are contingent, and that they are supported by expectations of others' compliance. Putting the fork to the left of the plate, wearing appropriate attire in social occasions, driving on the right side, and using a handkerchief to blow one's nose are all examples of patterns of behavior that might not have been. Furthermore, each such behavioral pattern elicits my conformity on the condition that I expect most others to conform to it. This suggests that a behavioral regularity would naturally be called a norm only if that regularity would not spontaneously arise by virtue of each individual's pursuing his own ends irrespective of his expectations about what other members of society would do.

Norms usually allow an individual to anticipate the behavior of others. We normally expect people to conform to norms, and we expect others to expect us to conform, too. A social norm depends for its existence on a cluster of expectations. Expectations, I want to argue, play a crucial role in sustaining a norm. Conformity to a social norm is not unconditional: It is rather a conditional choice based on expectations about other people's behavior and beliefs. One's interest in avoiding the negative consequences of transgression, as well as the feelings of shame and guilt that may accompany it, *reinforce* one's tendency to conform. But these are neither the sole, nor the ultimate, determinants of conformity. Reducing conformity to unconditional utility maximization overlooks the conditional element that characterizes norm-abiding behavior. Besides, approval and disapproval are sanctions that *presuppose* the existence of norms that everyone expects to be followed. Consider a community that abides by a strict norm of truth telling. A foreigner who, upon entering the community, systematically violates this norm will be met with hostility, if not utterly excluded from the group. But suppose a large group of liars makes its way into this small society. In time, people would cease to expect truthfulness on the part of others and would find no reason to be truthful themselves in a world overtaken by deception. Probably the truth-telling norm would cease to exist, since the strength of a norm lies in its being followed by almost all of the participants.

It may seem that most people's experience of conformity to a norm is beyond rational calculation. Compliance may look like a habit, thoughtless and automatic, or it may be driven by feelings of anxiety at the thought of what would happen if one transgresses the norm. Indeed, most sociologists and psychologists who have studied norms argue that norms are independent motivating factors. I believe, too, that upholding a norm is not a matter of conscious cost-benefit calculations; rather, people tend to repeat patterns of behavior that they have learned and, on average, work well in a variety of situations. Yet conformity to a norm may be rational and may be explained in terms of one's beliefs and desires, even though one does not conform out of a rational calculation. As David Lewis himself pointed out in his analysis of habits, a habit may be under an agent's rational control in the sense that should that habit ever cease to serve the agent's desires according to his beliefs, it would at once be overridden and abandoned (Lewis 1975a, 1975b). Similarly, an explanation in terms of norms does not compete with one in terms of expectations and preferences, since a norm persists precisely because of certain expectations and preferences: If I ever wanted to be different, or if I expected others to do something different, I would probably overcome the force of the norm. One is not constantly aware of one's beliefs, preferences, and desires, which I take to be dispositions to act in a certain way in the appropriate circumstances. What is required in a dispositional account of belief, preference, and desire is that such motives be ready to manifest themselves in the relevant circumstances. If somebody were to ask you now if you prefer a BMW car and one thousand dollars to a punch in the nose and one hundred thousand dollars, I do not know what your answer would be. Whatever option you would choose, it is likely that you would never have thought of it before; *you would not know*, for example, that you preferred the BMW and one thousand dollars until you were put in the condition to choose. Analogously, when conforming to a norm, one may be unaware of the expectations and preferences that underlie one's behavior, which become manifest only when they happen to be unfulfilled.

What sorts of preferences and expectations underlie the conditional choice to conform? A norm exists because everyone expects everyone else to conform, and everyone knows he is expected to conform, too, but expectations alone cannot motivate a choice. If my compliance is grounded on the expectation of almost universal compliance, it must be that I prefer to comply with the norm on condition that almost everyone else complies as well. When going to a dinner party, I do not wear sneakers, not simply because I expect everybody else to wear proper shoes, but because I also prefer to wear proper shoes if everybody else does. Note that I need not assume that the other guests at the dinner party also have conditional preferences. My belief that they will wear proper shoes may be grounded on the idea that they actively dislike sneakers, or that perhaps they are very traditional and not given to casual dressing. Of course, I might fear that disappointing their expectations will bring about contempt and thus have some additional good reason to wear proper shoes. But this is an independent, secondary reason. Preference for conformity and the belief that everyone else conforms to the norm are the main motivating forces.

The idea that expectations play a crucial role in sustaining norms agrees with experimental evidence (Dawes et al. 1977) that people's expectations of others' behavior matter in their choosing to contribute to the provision of a public good. Dawes et al. show that people's expectations of others are based on how they themselves have behaved in the past or are disposed to behave now. That is, people tend to use their own behavior as a cue in predicting the choices of other people. Typically, contributors expect a high rate of contributions, and, even when this expected high rate means that their contribution would be redundant, they still decide to contribute. A plausible explanation of this apparently irrational behavior is that the contributors comply with a norm of cooperation. They expect others to contribute, too, but this expectation does not induce them to defect. On the contrary, they prefer to contribute, given their expectation of high conformity to a cooperative norm.

I am now in a better position to explicate the notion of norm. Let R be a behavioral regularity in population P. Then more generally, R is a *social norm* if and only if R depends upon the beliefs and preferences of the members of P in the following way:

1: Almost every member of P prefers to conform to R on the condition (and only on the condition) that almost everyone else conforms, too.
2: Almost every member of P believes that almost every other member of P conforms to R.

Conditions (1)–(2) define a social norm as sustained by the beliefs and preferences of those who conform to it; they tell us that a social norm is an equilibrium in the game-theoretic sense of being a combination of strategies, one for each individual, such that each individual's strategy is a best reply to the others' strategies, were one to take them as given. Each prefers to conform on the condition that nearly everybody else conforms to the norm. Note that conditional preference does not require that conformity is a dominant strategy; if it were, then one would have a reason to conform independently of what other people were expected to do, in which case the equilibrium would not be called a social norm. Moreover, it would become impossible to distinguish a norm from a habit, or a moral imperative. So, even if all social norms are Nash equilibria, the converse is not true.

Conditions (1)–(2) reconcile norms with strategic behavior, but at a price: Characterizing social norms as equilibria spells out the conditions under which norms can be upheld but does not indicate how these conditions can be realized. Since social norms are standards of behavior that have come to be expected by a group or community in a particular social setting, to describe how expectations arise and become self-fulfilling is a crucial part of an explanation of how norms emerge.

LEARNING TO COOPERATE

The equilibrium definition of norms I have just provided does not make any distinction between a norm that is followed by relatively few people and a norm that is shared by an entire population. Examples of the first are all those regular patterns of behavior that evolve in families, among friends, or in small, cohesive groups, such as clubs and teams. The second type of norm is best illustrated by traffic rules, norms of etiquette, and all forms of racial or sexual discrimination. These latter are norms of cooperation, since they allow a large group, sometimes an entire population, to benefit from excluding some other group from certain activities of goods. What distinguishes the two types of norms is the process through which they come into existence. In both cases, individuals will form some beliefs about other individuals, and if enough individuals share the same beliefs, they will act in a way that will make their beliefs self-fulfilling. In both cases individuals will learn to detect behavioral patterns, but what learning involves will differ according to the size of the group. Even in the absence of communication, in a two-person repeated interaction there is scope for signaling and for experimenting with different actions. In large groups, on the other hand, one's actions go mostly undetected, and all that one observes is the aggregate behavior of the group, which is the sole predictor of future outcomes. The individual's influence on the group is marginal, so there is no point in signaling or experimenting.

Although a norm may emerge through learning in a two-person interaction, it may never spread to a population, and, if it does, the mechanism accounting for its diffusion is likely to be very different from that which explains its formation. Size will not matter much in those cases in which the passage from few to many individuals does not involve a change of incentives. To illustrate the point, take the case of neighborhood segregation; a white family may prefer to stay in a certain neighborhood as long as other white families stay, so that if everybody expects others to stay, there will be no incentive to leave. When a black family moves in, the immediate neighbors may take it as a sign that further changes in the racial balance will follow. This fear may induce them to move, thereby generating further worries in their immediate neighbors, who may also decide to relocate. This "snowball" effect is an example of spontaneous coordination: It takes the action of one or two individuals (families, in this case) to generate a collective outcome involving an entire population. The norm "do not live in a racially mixed neighborhood" is an example of a pattern of behavior that, once established among a few individuals, rapidly spreads to larger numbers through a mechanism of self-fulfilling expectations. Moreover, the greater the number of people who move away, the stronger becomes the incentive to move. With norms of cooperation, instead, the incentives to follow the norm are inversely proportional to the number of people involved. This is why it is so difficult to specify a plausible process that accounts for the formation and spread of norms, especially in those cases in which a norm of reciprocity is shared by an entire population.

Under which conditions will individuals who interact with each other for some time come to adopt a cooperative pattern of behavior? Let us assume for the time being that the players are not motivated by some other norm, and that they are boundedly rational. Bounded rationality can be variously interpreted as satisficing (Simon 1982), limited memory (Kalai and Stanford 1987; Neyman 1985), or limited computability (Binmore and Samuelson 1992). I shall briefly discuss both the case in which players have only very limited memory and strategic capabilities, and the case in which players have the capability of taking into account the effects of their own actions on other players' actions.

As an example, imagine two individuals engaged in a prisoner's dilemma–type game, which they know will be repeated a finite number of times. They do not know each other, nor do they have previous experience with this situation. These people are rational and know that joint cooperation is better than joint defection, but each has no idea what sort of player her opponent is. After each round of play, each learns how her opponent has played and adapts her subsequent choices to what has been learned. There are many ways a player can adapt, depending on such variables as memory, pattern-recognition capability, and the ability to take into account the effects of her own adjustments upon her opponent's play.

The simplest possible case of adaptation is that in which the players are "limited strategists," in that they simply adapt their choice to the action taken by the opponent on the preceding play. Skyrms (1990) calls these strategies *Markov strategies*, because at each stage a player chooses an action independently of the history of the game except for the immediately preceding action. Agents who use such memoryless strategies will not try to identify complex patterns of play, nor will they change their strategy in response to another player's moves. Each player will start by cooperating/defecting and will subsequently respond with cooperation/defection to the action taken by the opponent. There are eight possible adaptive rules the players may choose:

Rule	First play	Subsequent plays	
		If other played C	If other played D
1.	C	C	C
2.	C	C	D
3.	C	D	C
4.	C	D	D
5.	D	C	C
6.	D	C	D
7.	D	D	C
8.	D	D	D

Rule 1 is "unconditional"; rule 2 is "tit-for-tat"; rules 3, 4, 5, and 7 make little sense, since they do not clearly indicate either good will or an exploitative attitude; rather, 4 and 5 seem to indicate a contradictory attitude, whereas 3 and 7 just tell a player to do the opposite of what the other did before. Rule 6 would be adopted by a cautious player, ready to respond cooperatively to a cooperative opponent but unwilling to be exploited even once by a defector; rule 8 is "unconditional defection," which may be adopted by an overly pessimistic player. We suppose the players' choice of a rule to mirror their psychological propensities, and since we assume the players to be adaptive in a very limited way, we do not expect them to change their strategy in the course of play, since this option would entail far greater learning capabilities on their part.

Now suppose that two Markov players, A and B, are playing one hundred repetitions of the prisoner's dilemma game shown in Figure 1. The resulting supergame (Figure 2) has two equilibria in pure strategies: Either both players defect, or they both play tit-for-tat. It is unreasonable, however, to suppose that players will consider all possible rules as equally meaningful. In fact, rules 3, 4, 5, and 7 can be eliminated, since they present combinations of initial moves and adaptive responses that make little sense, and in such a limited adaptive situation they cannot be perceived as "signaling" some complex pattern of play. The players are left with rules 1, 2, 6, and 8 to choose from.[2] Since the players have four possible patterns of play to choose from, they face the following 4×4 game, in which each of the four rules is a strategy that will be played in the hundred repetitions of the above prisoner's dilemma, and the payoffs are, as before, the undiscounted sum of the payoffs each player obtains in each repetition of the game shown in Figure 3.

The supergame of Figure 3 has two equilibria: Either both players play a tit-for-tat strategy (rule 2), or they both always defect (rule 8). Note that this simple adaptive behavior may not allow one to learn the strategy of an

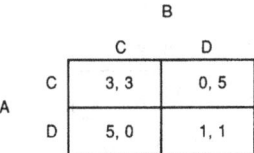

Figure 1.

		B	
		C	D
A	C	3, 3	0, 5
	D	5, 0	1, 1

	1	2	6	7	3	4	5	8
1	300, 300	300, 300	297, 302	0, 500	3, 498	3, 498	297, 302	0, 500
2	300, 300	300, 300	250, 250	225, 225	225, 225	101, 106	299, 299	99, 104
6	302, 297	250, 250	100, 100	225, 225	225, 225	103, 103	300, 295	100, 100
7	500, 0	225, 225	225, 225	200, 200	500, 0	6, 491	494, 4	1, 496
3	498, 3	225, 225	225, 225	0, 500	200, 200	4, 494	493, 8	0, 500
4	498, 3	106, 101	103, 103	491, 6	494, 4	102, 102	495, 5	99, 104
5	302, 297	299, 299	295, 300	4, 494	8, 493	5, 495	298, 298	1, 496
8	500, 0	104, 99	100, 100	496, 1	500, 0	104, 99	496, 1	100, 100

Figure 2.

		B			
		1	2	6	8
A	1	300, 300	300, 300	297, 302	0, 500
	2	300, 300	300, 300	250, 250	99, 104
	6	302, 297	250, 250	100, 100	100, 100
	8	500, 0	104, 99	100, 100	100, 100

Figure 3.

31

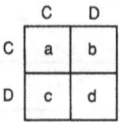

Figure 4.

opponent; a tit-for-tat player will never learn whether she was matched with an unconditional cooperator or a tit-for-tatter, and an uncompromising defector will never know whether she was matched with a cautious, cooperative type. Even if learning were to occur, it would not be exploited to the advantage of the players: A prudent cooperator adopting rule 6 will immediately learn whether her opponent is an unconditional cooperator but will not exploit this knowledge to her advantage, while a prudent cooperator and a tit-for-tat player will "lock" into a punitive pattern even if, by being lenient just once, either one could induce a dramatic improvement in the overall outcome.

Consider what a Markov player would choose, knowing that she can only marginally adapt. If the opponent is a conditional cooperator, it is better to cooperate, while a tit-for-tat strategy does little harm to a player in case her opponent is the "always defect" type. If matched to a prudent cooperator, it would be better to be unconditionally cooperative, but unconditional coopera-tion is too risky a prospect, while tit-for-tat still does better than the remaining rules. Note that tit-for-tat is a better prospect even if the opponent is "smarter" than the player. Tit-for-tat will protect the player from exploitation by someone endowed with greater learning capabilities, since her defection will be immedi-ately punished. Hence we would expect adaptive players to choose rule 2 and settle on the (2, 2) equilibrium.

This result, however, crucially depends on the payoffs associated with the outcomes. Consider the following prisoner's dilemma matrix, in which the letters represent the payoffs obtained by the row player for each combination of her and the opponent's strategies (Figure 4).

Since the game in Figure 4 is a prisoner's dilemma, $c > a > d > b$.[3] For n plays of the game, dn is what a defector will score if matched with another defector, while a tit-for-tat player matched with a defector gets $d(n-1) + b$. If $dn - (d(n-1) + b)$ is small, a player will be willing to cooperate, but as $d(n-1) + b$ decreases, the probability of choosing defection increases. If unilateral cooperation were associated with a large loss, defection would be more likely to obtain. Note that a player need not assign a high probability to being matched with a cooperator in order to choose a tit-for-tat strategy. In fact, in one hundred repetitions of the game in Figure 1, it is a sufficient to assess probability 0.047 that the other player is a tit-for-tatter in order to choose tit for tat. In general, the greater the loss associated with unilateral cooperation, the higher the probability of being matched with a cooperator must be in order to make one choose to be a conditional cooperator.[4]

Suppose now that the rewards associated with joint cooperation and the punishments associated with joint defection are large enough to justify a cooperative choice in a two-player repeated game, but that the game being played is an n-person prisoner's dilemma. If n is large enough to guarantee a player's anonymity, the incentives change and universal defection is to be expected. If instead the number of players is such that one's defection can be easily detected and punished, the cooperative equilibrium remains a possibility. The choice to be a conditional cooperator will still depend upon whether unilateral cooperation is not too costly, and, if the payoff structure is favorable enough, individual cooperation can be expected to continue as long as the group's past aggregate behavior is cooperative. A behavioral regularity thus established can be highly unstable, though. For one, it will be very sensitive to variations in the payoff structure. For example, since the players' adaptive rule only considers what happened in the previous play, the fact that everyone has "conformed" to a cooperative pattern for a long time has no effect upon the choice of a rule the next time a prisoner's dilemma situation occurs: With a payoff structure unfavorable to cooperation, an individual would choose to defect. This conclusion is in line with what has been observed by other writers in the field. Russel Hardin (1982) has pointed out that often a cooperative outcome in the context of a repeated prisoner's dilemma is due to the existence of "extra-play" incentives that influence current choices. Such incentives, however, require a certain degree of sophistication on the part of the players.[5] An example of an extra-play incentive is the prospect of future activities involving the same participants, or different ones who will be informed about the past behavior of the current players. Under these circumstances, it will be in the individual's interest to create a reputation for being a trustworthy, cooperative type. But reputation effects require that a player be able to evaluate the future consequences of present behavior.

What we need is a more complex adaptive dynamics in which strategic uncertainty and the acknowledgment of the possible effects of one's adjustments on other players' adjustments play a greater role. Consider again two rational players engaged in a prisoner's dilemma with a known number of repetitions who do not have any information about their opponent's type. The players may introspect to form their prior probabilities about the opponent's type, and in subsequent play try to test those different hypotheses. For example, two players may start by cooperating and see what happens next. After a few repetitions, both will have formed some hypothesis about their opponent's strategy and will adjust their strategies accordingly. In a simple, two-person interaction, it is likely that a player will take into account the effects of her own adjustments upon her opponent's play, whereas in a large population one will take the current state of the rest of the population as a prediction of its state at the next stage, since the effects of one's adjustments on other people's future adjustments are insignificantly small.

In order to eliminate some hypotheses about the opponent's type, the players may "experiment" with small deviations, which may or may not be profitable in terms of payoffs (Fudenberg and Kreps 1988). Consider for example the case in which both players cooperate for n repetitions. Then a player may want to "test" the hypothesis that her opponent is a retaliator by defecting the next round. Her deviation reveals to the other player that she is not an unconditional cooperator, but it might also be taken as a signal that she is unwilling to cooperate from now on, if the number of repetitions is small and the end of the game is not too far away. Testing a hypothesis involves deviating from one's chosen strategy, and it involves the risk of being misperceived by another player as playing a different strategy. Taking this possibility into account, a rational player will be more likely to experiment at the beginning of the game, in order to restrict the set of possible conjectures about her opponent's type without "confusing" her too much.

Thus a tit-for-tat player may want to ascertain that she is not playing with an unconditional cooperator, and a defector may want to test the willingness of her opponent to "forgive," as well as the severity of her retaliation policy. A prudent cooperator may want to ascertain whether she is playing with another prudent cooperator instead of a defector, whereas an individual who believes she "deserves more" and thus plans to defect, say, twice every three plays, will want to know whether she is playing with a similarly convinced player or with a tit-for-tat retaliator. In the latter case, the two patterns of play may look identical:

> Player 1. C D D C D D C D D C D D . . .
> Player 2. D D C D D C D D C D D C . . .

Player 1 may be a tit-for-tat player, but she may also be a conditional cooperator who, like player 2, believes she is more deserving than her opponent. If player 2 initially defects, it will be impossible for her to know whether or not player 1's pattern of play is independent of her choices. A better strategy for player 2 would be to initially cooperate, since by the fifth play she will know whether she is playing against a retaliator, in which case she would do better by modifying her strategy and choose to cooperate until the penultimate play. To see why this is so, consider the following pattern of play of player 2 (who believes she deserves more) against a tit-for-tat player:

> Player 1. C C D D C D . . .
> Player 2. C D D C D D . . .

The first three moves of player 1 may suggest a pattern of limited cooperation that tells a player to defect once every three plays, the fourth move may or may not indicate retaliation, but by the fifth move player 2 may come to see that player 1's next move is always identical to her own previous move, signaling a retaliator who is quick to "forgive" defection and reward cooperation. Note that "tougher" retaliatory strategies will not do as well as tit-for-tat. For example, a tougher retaliator may delay rewarding the opponent for cooperative behavior if there has been a previous defection by restoring cooperation only after the opponent has unsuccessfully cooperated once. The pattern of play may look like this:

Player 1. C D D D D D C...

Player 2. D D C D D C...

Suppose player 2 is an exploitative type who would always defect in the presence of unconditional cooperators, unless she is convinced that she is playing with a conditional cooperator who will punish her defection and reward her cooperation, in which case she would maximize her expected utility by cooperating. Since player 1 chooses to cooperate in the first play and defect in the second, player 2 suspects she is a tit-for-tat player; in order to test her hypothesis, player 2 may deviate from her strategy in the third play, since a tit-for-tat player would respond positively to her signal in the next play. If player 1 keeps defecting, player 2 will know for sure that she is not a tit-for-tat player, but the set of possible strategies she may be playing is still very large. Player 1 may be a conditional cooperator who punishes defection by defecting forever after; she may be willing to signal a cooperative attitude at fixed intervals; she may be a defector who made a mistake in her first move; she may be a tough retaliator who will exploit twice or more in return for each exploitative episode she had to suffer, and so on. Depending on the projected costs of undergoing further testing, which will depend both on the assessed probabilities of each hypothesis and the magnitude of the loss associated with unreciprocated cooperation, player 2 may or may not attempt further testing. This example suggests that a tough retaliator risks locking herself in a self-defeating pattern, since punishment, to be effective, must be easy to understand.[6]

Prudent cooperation is a difficult strategy to detect, too, since the prudent cooperator will initiate the game by defecting and will subsequently cooperate in response to a cooperative move of his opponent. When two prudent cooperators are matched, their strategies will be indistinguishable from an "always defect" strategy, and they will keep defecting unless one of them is willing to risk being twice "exploited" in order to test the hypothesis that the opponent is not a defector.

In general tit-for-tat has a big advantage over other strategies: It is easy to learn, since it has a clearly recognizable pattern, and it protects the player who adopts it from excessive exploitation by a defector; tit-for-tat will at best tie, at worst it will lose no more than one play. Robert Axelrod (1984) contends that these features of tit-for-tat are responsible for his overwhelming success in the computer tournaments he ran. When different strategies were paired off for round-robin tournaments of an iterated prisoner's dilemma, Axelrod found out that if the probability of the game's continuing was sufficiently great, tit-for-tat scored better than all the other strategies it competed with. The very characteristics that make tit-for-tat successful in computer tournaments are also likely to play an important role in all those prisoner's dilemma–like circumstances in which there is repeated interaction but the players are uncertain as to the opponent's character.

However, in Axelrod's model there is no explanation for why individuals come to hold certain strategies originally other than by chance. In his evolutionary model, initial strategies are randomly chosen from the set of all possible strategies. I believe that, on the contrary, there are cognitive constraints on the possible strategies that may emerge as stable behavioral patterns in repeated, small-group interactions. For example, a hypothesis that should be tested is whether, when the number of repetitions is small and experimenting more costly, tit-for-tat is more likely to be chosen by a player who wants unambiguously to signal his intentions and benefit from the possibility of joint cooperation.

FROM SMALL TO LARGE GROUPS

Once a cooperative equilibrium is established, we may expect it to persist, since data from past experience can be used to predict how an opponent will act in the future. If we learn that we are playing with a tit-for-tat opponent, we recognize that unilateral defection is going to be immediately punished. Under these circumstances, each player will prefer to cooperate if the other cooperates, and each will attach high probability to the opponent's playing his part in the equilibrium; hence, each player will have a decisive reason to stick to cooperative behavior. Note that common knowledge has not been assumed, nor is it indeed to maintain conformity. Since the players will have probabilistic beliefs "close" to the equilibrium, but not full knowledge, beliefs will be quasi-consistent but not necessarily fully consistent with each other (Fudenberg and Kreps 1988). However, once players' expectations are close to the cooperative equilibrium, they will tend to persist because of reinforcing feedback: Each player will play his part in the equilibrium, which will lead the other to expect with greater certainty that the equilibrium accurately predicts what the opponent will do in the future.

Such a stable equilibrium is a norm of cooperation, since it fulfills the conditions that define a social norm as a function of the preferences and beliefs of the members of the population in which that behavioral regularity exists. It is easy to see how approval and disapproval play only a secondary role in eliciting conformity. If another player defects, one is made worse off in two respects: There is an immediate loss, and one is forced to punish the defector in the next round. The obvious disapproval that accompanies defection reinforces a cooperative attitude; it cannot, however, substitute for conditional preferences and beliefs in eliciting cooperation. We may expect a norm of cooperation to emerge as a stationary equilibrium in a group of players in which the identities of the players and the experience they have had with each other matter. Once a norm of cooperation has been established in a small-group interaction, it will tend to persist and elicit conformity in new situations in which both cooperative and competitive strategies are possible. If the subjects involved are the same, or if they carry with them reputations from past play, mutual expectations are likely to be quasi-consistent (Fudenberg and Kreps 1988), since each individual will tend to believe that what has happened in the past is a good predictor of what will happen in the future.

The larger the population becomes, however, the more individuals will tend to ignore the effects of their adjustments on the future course of play of other individuals, as their identity (and reputation) will matter less and less. In fact, I doubt that learning is possible in large, anonymous groups. In large groups, an individual's choice has an insignificant impact on the collective outcome, and defection is likely to go undetected. In those circumstances, experimenting with small deviations from one's strategy is pointless, since no response is likely to follow. The only data available to predict the future state of the population are its past and current aggregate behavior, so if cooperation has taken place in the past, individuals will tend to expect it to occur in the future, too. In these circumstances, a game theorist would predict that expectations of cooperative behavior will be self-defeating, and expectations of defecting behavior self-fulfilling. However, experiments unequivocally show that a large proportion of subjects tend to cooperate even in one-shot situations of complete anonymity. Moreover, Dawes et al. (1977) found out that subjects who expect a high degree of cooperation tend to cooperate, even if they are aware that their contribution is redundant. Existing models of rational Nash-equilibrium play cannot explain these results. To understand the above anomalies, a two-tiered model of how norms of cooperation emerge and spread (or decay) is in order.

If people can learn to cooperate only in dyadic or small-group interactions, the explanation of how norms of cooperation emerge as equilibrium patterns of behavior does not extend to large, anonymous groups, in which the presence of conforming behavior might be better explained by the diffusion of small-group norms through an evolutionary process. Russel Hardin (1982) has pointed to the overlapping nature of group activities and the tendency to generalize to similar cases as possible mechanisms through which conventions involving large populations are built up out of dyadic interactions. Examples are the norms of truth telling and promise keeping. One will presumably learn that it is better to be sincere and trustworthy in the context of repeated interactions with the same small group of people and will later adopt the same behavior in situations that are perceived as sufficiently similar to the original ones or that involve reputation effects, in that violation of the norm is taken to signal a flawed character, and this in turn will jeopardize future interactions.

Another striking example of widespread cooperative behavior is the development of the "honor code" that governed international commerce in the thirteenth century. It was common for merchants to buy on credit and clear their debts at some later time; all the seller got was a "bill of exchange," a written promise to pay a sum of money at some later date. Henri Pirenne notes that since metallic money was scarce, the massive development of commerce was made possible by the practice of credit, which involved the use of bills of exchange as money.[7] It was not the enforcement of a government but rather the trustworthiness of the issuer that backed a bill of exchange and made it usable as a means of payment. All this would have been unfeasible had the merchants not been in continual relations of debt and credit with one another, and thus concerned with their good standing among their peers. One can imagine the original development of norms of business conduct among a few local merchants, their spread to a larger business community through the repeated contacts with foreign merchants provided by international fairs, and finally the emergence of a general, unspoken code of behavior regulating the activities of an international community of merchants and bankers.

I believe that an analysis of how norms are applied and maintained must involve a cognitive element. Recent empirical work on repeated PD games indicate that individuals appear to be using precedent-based reasoning in solving new decision problems: People tend to select past situations most closely resembling the present situation and apply the prescriptions of the past to the present (Alker and Hurwitz 1980; Alker et al. 1980). This work is consistent with bounded rationality approaches in cognitive psychology (Winson 1980) and artificial intelligence (Newell and Simon 1972; Carbonell 1983) that stress the importance of precedent-based reasoning and analogical inference in decision making.

However, if the maintenance of norms has a cognitive component, it is also a social process. Once a cooperative pattern of behavior has been established, people will tend to expect it to persist and use cooperative norms as "default values" in new social contexts. But the social environments in which people find themselves are crucial in determining the survival or demise of social norms. In this context, an evolutionary approach is useful in studying the dynamics of norms, because it can show how strategies change over time as a function of their relative success in an ever-changing environment. For example, though an established norm of cooperation has motivational power, people who usually uphold it will tend to abandon it if it becomes evident that it does not do as well as other patterns of behavior in the new environment. Indeed, Dawes et al. (1977) have noticed that, although cooperation is high in the initial stages of a repeated public-good game, it tends to decay rapidly whenever the subjects realize that they have been "cheated" by others.

An evolutionary model of the spread of a behavioral pattern over larger groups or even entire populations is not in conflict with an explanation of its emergence in terms of individual learning in repeated, small-group interactions. Voting, giving to charities, refraining from littering or polluting are choices that are not easily amenable to a rational explanation. They need not, however, be thought of as counterexamples to rational choice theory. My thesis is that they result from compliance with norms of cooperation that emerged in other contexts out of self-interested behavior and were subsequently extended to the entire population through pressures that are analogues to selection in the wild. The advantages of supplementing a rational choice explanation with an evolutionary approach are twofold. On the one hand, an evolutionary model does not require sophisticated reasoning and learning in circumstances, such as large-group interactions, in which it would be unrealistic to assume them. We may rather suppose that some behavioral patterns borne of strategic interactions spread and evolve in a large population out of simple adaptive mechanisms. It is not too far-fetched to assume that strategies that make a person do better than others will be retained, while strategies that lead to failure will be abandoned. Another plausible mechanism is imitation: Those who do best are observed by others who subsequently emulate their behavior (Bicchieri and Rovelli 1995).

Whether a behavioral pattern that has emerged in a small group will survive in a larger population is an important question to address, and an evolutionary model provides a description of the conditions under which social norms may spread. One may think of several environments to start with. A population can be represented as entirely homogeneous, in the sense that everybody is adopting the same type of behavior, or heterogeneous to various degrees. In the former case, it is important to know whether the commonly adopted behavior is stable against mutations. The relevant concept here is that of an *evolutionarily stable strategy* (Maynard Smith and Price 1973; Taylor and Jonker 1978); when a population of individuals adopts such a strategy, it cannot be successfully invaded by isolated mutants, since the mutants will be at a disadvantage with respect to reproductive success. A more interesting case, and one relevant to a study of the reproduction of norms of cooperation, is that of a population in which several competing strategies are present at any given time. What we want to know is whether the strategy frequencies that exist at a time are stable, or if there is a tendency for one strategy to become dominant over time.

There are important differences between the model presented here and other evolutionary models of the emergence of cooperation. Kin-selection models (Hamilton 1964), being gene-centered, cannot explain cooperative behavior toward strangers, whereas the model I propose succeeds in explaining under which conditions such cooperative behavior can emerge and spread. Models of "reciprocal altruism" (Trivers 1971, 1985), on the other hand, tell us that cooperative behavior has no chance of evolving in random pairings but will evolve in a social framework in which individuals can benefit from building "reputations" for being cooperative types. Reciprocal altruism, however, does not require an evolutionary argument; a simple model of learning in small-group interactions such as the one I propose will do, and has the further advantage of explaining why certain types of cooperative behavior are more likely to emerge than others. Finally, my model differs from Axelrod's (1984) in that in my model the set of initial strategies is not arbitrary. Since I argue that norms of cooperation emerge in small-group interactions among boundedly rational agents, I can predict that the patterns of behavior that emerge as norms will generally be simple ones (i.e., they can be modeled as finite automata with a small number of states). If this is the case, the set of strategies that are present at any given time in a population is bounded by complexity considerations. Moreover, different strategies might be used by different fractions of the original population, and the resulting equilibrium will depend on the original strategy distribution.

What follows is a simple example of an evolutionary process (Bicchieri 1990). The dynamics I model have the property that strategies with higher current payoffs increase in frequency over time. A game is repeated N times, and after each round of the game, the actual payoffs and strategies of the players become public knowledge; on the basis of this information, each player is allowed to adjust his strategy for the next round. More formally, let P_{it} be the frequency of strategy i in population P at time t, and let Π_{ij} be the payoff to adopting strategy i if the opponent plays strategy j. Let

$$\Pi_{it} = \sum_j \Pi_{ij} p_{it}$$

be the total payoff of playing i at time t, which may also be interpreted as i's fitness at time t. Note that the total payoff is the weighted sum of the different payoffs one obtains by being matched with different types of strategies, where the weights represent the frequencies of those strategies in the entire population at time t. $p_{it+1}(p_{1t}, p_{2t}, \ldots)$ represents the frequency of strategy i at time $t+1$ as a function of the relative frequencies at time t of all the available strategies (including i), and thus depends on the total payoff of playing strategy i at time t, since it is the payoff one obtains by playing a given strategy that determines whether one is going to play it again or abandon it. The dynamics of the frequency distributions of the strategies can be represented as follows:

$$(1) \qquad p_{it+1} = \frac{f(\Pi_{it})p_{it}}{\sum_j f(\Pi_{it})p_{it}},$$

where $f(\Pi_{it})$ is the reproduction rate of strategy i and is a monotonically increasing function of the total payoff of playing strategy i at time t.

An equilibrium is a frequency vector $(p_1, p_2, \ldots)$ that reproduces itself over time. A pure strategy equilibrium is one in which only one strategy is played by the entire population; that is, it is a vector where $p_n = 1$, and $p_i = 0$ for all $i \neq n$. Of course, any such vector reproduce itself, but what we want to know is whether it is stable against mutations. Given two strategies i and n, the frequency of i over time will decrease if the reproduction rate of strategy n matched with n is greater than that of strategy i matched with n, that is, $p_{i1} > p_{it+1}$ in each period as long as the ratio

$$\frac{f(\Pi_{in})}{f(\Pi_{nn})} < 1 \text{ for each } i \neq n.$$

Given a fixed population, the number of mutants rises whenever the fitness associated with the mutant strategy is higher. Following Maynard Smith (1982), the condition that makes an equilibrium stable over mutations is that, for each $i \neq n$, $\Pi_{in} < \Pi_{nn}$.

Consider the simple case in which there are only two possible strategies: tit-for-tat (T) and defect (D). This simplification helps in understanding the dynamic process, which becomes much more complex with a larger number of strategies, but the simplification does not affect the generality of the analysis.

Let $p = $ frequency of strategy T (tit-for-tat) and $1 - p = $ frequency of strategy D (defect). We may rewrite equation (1) as

$$p_{t+1} = \frac{f(\Pi_{Tt})p_t}{f(\Pi_{Tt})p_t + f(\Pi_{Dt})(1 - p_t)}$$

$$= \frac{f(p_t\Pi_{TT} + (1 - p_t)\Pi_{TD})p_t}{f(p_t\Pi_{TT} + (1 - p_t)\Pi_{TD})p_t + f(p_t\Pi_{DT} + (1 - p_t)\Pi_{DD})(1 - p_t)}.$$

We want to find the solution $\hat{p}$ to the above equation in the nontrivial case in which $\hat{p}$ is different from 0 and 1. That is, we look for a value of p such that, if the frequency of tit-for-tatters in the population is equal to or greater than it, the dynamics will favor tit-for-tatters and, at that value, the number of conditional cooperators will be stable. Substituting $\hat{p}$ in the above equation, we obtain

$$f(\hat{p}\Pi_{TT} + (1 - \hat{p})\Pi_{TD}) = f(\hat{p}\Pi_{TT} + (1 - \hat{p})\Pi_{TD})\hat{p}$$
$$+ f(\hat{p}\Pi_{DT} + (1 - \hat{p})\Pi_{DD})(1 - \hat{p})$$

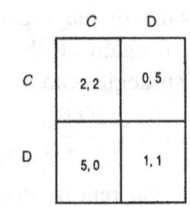

Figure 5.

and solving for $\hat{p}$, we get

$$\hat{p} = \frac{\Pi_{DD} - \Pi_{TD}}{\Pi_{TT} + \Pi_{DD} - \Pi_{DT} - \Pi_{TD}}.$$

Suppose there have been N repetitions of the game shown in Figure 5. The payoffs obtained by playing T or D against an opponent who plays, respectively, T or D are the following:

$$\Pi_{DD} = N; \quad \Pi_{TT} = 2N; \quad \Pi_{TD} = N - 1; \quad \Pi_{DT} = N + 4.$$

In this case, assuming that N is at least 4, the minimum value at which a group of tit-for-tatters can survive in a population of defectors is

$$\hat{p} = \frac{N - (N - 1)}{2N + N - (N + 4) - (N - 1)} = \frac{1}{N - 3}.$$

If N is greater than 4, this is a fraction less than one. For values greater than $\hat{p}$, tit-for-tat will not just survive but thrive, as more and more players will see the advantage of adopting it. However, if the frequency of tit-for-tatters is lower than $\hat{p}$, cooperative behavior will get less and less frequent in each period and will eventually disappear (Figure 6). On the other hand, if N is less than 4, no matter how many tit-for-tat mutants enter a population of defectors at once, their frequency will decrease in each period and eventually go to zero (Figure 7).

This simple example illustrates the general point that norms of cooperation that emerged in a small group may extend to a population through an evolutionary process; if the number of repetitions is sufficiently large, a small proportion of cooperators can take over an entire population (Bicchieri and Rovelli 1995).

Note, however, that the evolutionary stability of a norm of cooperation will depend on the presence or absence of other strategies in the environment, as well as on the hypothesis that certain strategies are permanently wiped out. Suppose for example that in a population of defectors (D), unconditional co-operators (C), and tit-for-tatters (T), the defectors die out before the C-players do. In the new environment, C can become as fit as T, since there are no defectors around to punish the naivete of the C-players. And in the protracted

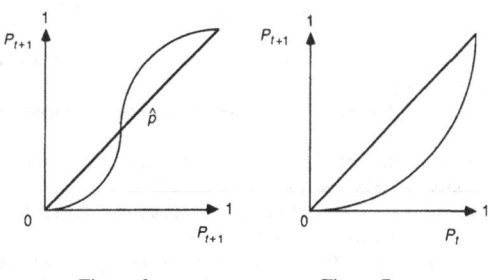

Figure 6.   Figure 7.

absence of selection pressures against C-players, they may come to replace the
T-players. In this event, immigration of a new group of D-players will wipe out
the cooperators (Young and Foster 1991). The sensitivity of the dynamics of
norms to initial conditions, such as the type and proportions of norms that are
present in a population at any given time, make the study of their emergence
particularly relevant. If norms can emerge only through learning in small-group
interactions, then the cognitive constraints of the learning agents will impose
bounds on the set of norms that can possibly emerge. As the initial conditions
of the evolutionary models become less arbitrary, we can make more reliable
predictions as to the evolution of norms of cooperation.

NOTES

1. This result is proved in Kreps et al. (1982).
2. Brian Skyrms has discussed these rules, which he calls "Markov habits," in the
   context of dynamic deliberation on the part of Bayesian players. If Bayesian dynamic deliberators have adequate common knowledge for deliberation, the greater
   the number of iterations, the greater the degree of mutual distrust needed to justify
   the selection of the "always defect" rule. Cf. B. Skyrms (1990, ch. 6.)
3. It must also be the case that $b + c < 2d$. And for $p \in [0, 1]$, $(pb + (1 - p)c) < d$,
   since otherwise players would use mixed strategies alternating between the cells
   (C, D) and (D, C) that are better for both than the payoff obtained from playing (D,
   D).
4. This probability will remain quite small, though. For example, if the payoffs of the
   game in Figure 1 are slightly modified so that the loss associated with unconditional
   cooperation is -5, the probability of being matched with a tit-for-tatter must be at
   least 0.069 for a player to choose to play tit-for-tat in one hundred repetitions of that
   game.
5. R. Hardin (1982, p. 164).
6. I suspect that, whenever an unfair pattern of cooperation emerges, this is more likely
   to be due to an underlying bargaining game in which one of the parties has greater
   bargaining power rather than due to poor learning.
7. H. Pirenne (1936).

43

References

Akerlof, G. 1976. "The Economics of Caste and of the Rat Race and Other Woeful Tales." *Quarterly Journal of Economics* **90**:599–617.

Alker, H. and R. Hurwitz. 1980. *Resolving Prisoner's Dilemmas*. American Political Science Association.

J. Bennett, and D. Mefford. 1980. "Generalized Precedent Logics for Resolving Insecurity Dilemmas." *International Interactions* **7**:165–206.

Anderson, P. 1981. "Justification and Precedents as Constraints in Foreign Policy Decision Making." *American Journal of Political Science* **25**:738–61.

Andreoni, J. and J. Miller. 1991. "Rational Cooperation in the Finitely Repeated Prisoner's Dilemma: Experimental Evidence." Working paper, Carnegie Mellon University.

Arrow, K. 1971. "Political and Economic Evaluation of Social Effects and Externalities." In M. Intriligator, ed., *Frontiers of Quantitative Economics*. North Holland.

Axelrod, R. 1984. *The Evolution of Cooperation*. New York: Basic Book.

1986. "An Evolutionary Approach to Norms." *American Political Science Review* **80**:1095–1111.

Bicchieri, C. 1989. "Self-Refuting Theories of Strategic Interaction: A Paradox of Common Knowledge." *Erkenntnis* **30**:69–85.

1990. "Norms of Cooperation." *Ethics* **100**:838–61.

1992. "Knowledge-Dependent Games: Backward Induction." In C. Bicchieri and M. L. Dalla Chiara, eds. *Knowledge, Belief, and Strategic Interactions*. Cambridge University Press.

Bicchieri, C. and C. Rovelli. 1995. "Evolution and Revolution: The Dynamics of Corruption." *Rationality and Society* **7**:201–24.

Binmore, K. and L. Samuelson. 1992. "Evolutinary Stability in Repeated Games Played by Finite Automata." *Journal of Economic Theory* **57**:278–305.

Blau, P. 1964. *Exchange and Power in Social Life*. Wiley.

Boyd, R. and P. Richerson. 1985. *Culture and the Evolutionary Process*. University of Chicago Press.

Camic, C. 1986. "The Matter of Habit." *American Journal of Sociology* **91**:1039–87.

Cancian, F. 1975. *What Are Norms?* Cambridge University Press.

Caporael, L., R. Dawes, J. Orbell, and A. van de Kragt. 1989. "Selfishness Examined: Cooperation in the Absence of Egoistic Incentives." *Behavioral and Brain Sciences* **12**:683–739.

Carbonell, J. 1983. "Learning by Analogy: Formulating and Generalizing Plans from Past Experience." In R. Michalski, J. Carbonell and T. Michell, eds., *Machine Learning: An Artificial Intelligence Approach*. Togia Publishing.

Coleman, J. 1989. *Foundations of Social Theory*. Harvard University Press.

Cooper, R., D. DeJong, R. Forsythe, and T. Ross. 1990. "Cooperation without Reputation." Working paper, University of Iowa.

Crawford, V. 1989. "Learning and Mixed Strategy Equilibria in Evolutionary Games." *Journal of Evolutionary Biology* **140**:537–50.

Crawford, V. and H. Haller. 1990. "Learning How to Cooperate: Optimal Play in Repeated Coordination Games." *Econometrica* **58**:571–95.

Dawes, R. 1980. "Social Dilemmas." *Annual Review of Psychology* **31**:169–93.

J. McTavish, and H. Shaklee. 1977. "Behavior, Communication, and Assumptions about Other People's Behavior in a Commons Dilemma Situation." *Journal of Personality and Social Psychology* **35**:1–11.

and R. Thaler. 1988. "Anomalies: Cooperation." *Journal of Economic Perspectives* **2**:187–97.

Diamond, A. 1935. *Primitive Law*. Watts.

Durkheim, E. 1950. *The Rules of Sociological Method*. Free Press.

Elster, J. 1989. *The Cement of Society*. Cambridge University Press.

Foster, D. and P. Young. 1990. "Stochastic Evolutionary Game Dynamics." *Theoretical Population Biology* **38**:219–32.

Fudenberg, D. and E. Maskin. 1986. "The Folk Theorem with Discounting and with Incomplete Information." *Econometrica* **54**:533–54.

and D. Kreps. 1988. *A Theory of Learning, Experimentation, and Equilibrium in Games*. Mimeo, Stanford University.

Gauthier, D. P. 1975. "Coordination." *Dialogue* **14**:195–221.

Geertz, C. 1973. *The Interpretation of Cultures*. Basic Books.

Gibbs, J. 1965. "Norms: The Problem of Definition and Classification." *American Journal of Sociology* **70**:586–94.

1981. *Norms Deviance, and Social Control*. North Holland.

Gilbert, M. 1981. "Game Theory and Conventions." *Synthese* **46**:41–93.

1983. "Agreements, Conventions, and Language." *Synthese* **54**:375–404.

1990. "Rationality, Coordination and Convention." *Synthese* **84**:1–21.

Gouldner, A. 1960. "The Norm of Reciprocity: A Preliminary Statement." *American Sociological Review* **25**:161–78.

Hamilton, W. 1964. "The Genetical Evolution of Social Behavior." *Journal of Theoretical Biology* **7**:1–52.

Hardin, R. 1982. *Collective Action*. Johns Hopkins University Press.

Hoebel, E. 1954. *The Law of Primitive Man*. Harvard University Press.

Homans, G. 1961. *Social Behavior: Its Elementary Forms*. Harcourt.

Jordan, J. 1991. "Bayesian Learning in Normal Form Games." *Games and Economic Behavior* **3**:82–100.

Kalai, E. and E. Lehrer. 1990. "Rational Learning Leads to Nash Equilibrium." Working paper, Northwestern University.

and W. Stanford. 1987. "Finite Rationality and Interpersonal Complexity in Finitely Repeated Games." *Econometrica* **56**:397–410.

Kandori, M. 1989. "Social Norms and Community Enforcement." Working paper, Stanford University.

Kreps, D. 1990. *Game Theory and Economic Modeling*. Oxford University Press.

P. Milgrom, J. Roberts and R. Wilson. 1982. "Rational Cooperation in the Finitely Repeated Prisoner's Dilemma." *Journal of Economic Theory* **27**:245–52.

Lewis, D. 1969. *Convention*. Cambridge: Harvard University Press.

1975. "Languages and Language." In K. Gunderson, ed., *Minnesota Studies in the Philosophy of Science*. Vol. 3. Minneapolis: University of Minnesota Press.

1975. "Convention: Reply to Jamieson." *Canadian Journal of Philosophy* **6**:1.

J. Maynard Smith. 1982. *Evolution and the Theory of Games*. Cambridge: Cambridge University Press.

and G. Price. 1973. "The Logic of Animal Conflict." *Nature* **246**:15–18.

Milgrom, P. and J. Roberts. 1991. "Adaptive and Sophisticated Learning in Normal Form Games." *Games and Economic Behavior* **3**:82–100.

Miller, S. 1990. "Rationalizing Conventions." *Synthese* **84**:23–41.

Nachbar, J. 1990. "Evolutionary Selection Dynamics in Games: Convergence and Limit Properties." *International Journal of Game Theory* **19**:59–89.

Newell, A. and H. Simon. 1972. *Human Problem Solving*. Englewood Cliffs.

Neyman, A. 1985. "Bounded Complexity Justifies Cooperation in the Finitely Repeated Prisoner's Dilemma." Unpublished manuscript.

Opp, K. 1979. "Emergence and Effects of Social Norms." *Kylos* **32**:775–801.

1983. "Evolutionary Emergence of Norms." *British Journal of Social Psychology* **21**:139–49.

Pancer, M. 1973. *Approval Motivation in the Prisoner's Dilemma Game*. Mimeo, University of Toronto.

Parsons, T. 1937. *The Structure of Social Action*. Free Press.

1951. *The Social System*. Free Press.

Piliavin, J. and H. Charng. 1990. "Altruism: A Review of Recent Theory and Research." *Annual Review of Sociology* **16**:27–65.

Pirenne, H. 1936. *Economic and Social History of Modern Europe*. Routledge and Kegan Paul.

Rapport, A. 1988. "Experiments with N-Persons Social Traps I. Prisoner's Dilemma,. Weak Prisoner's Dilemma, Volunteers, Dilemma, and Largest Numbers." *Journal of Conflict Resolution* **32**:457–72.

—— and A. Chammah. 1965. *Prisoner's Dilemma*. University of Michigan Press.

Rommetveit, R. 1955. *Social Norms and Roles*. University of Minnesota Press.

Schelling, T. 1960. *The Strategy of Conflict*. Harvard University Press.

—— 1978. *Micromotives and Macrobehavior*. Norton.

Schrodt, P. 1985. "Precedent-Based Logic and Rational Choice: A Comparison." In M. Ward and U. Luterbacher, eds., *Dynamic Models of International Conflict*. Lynne Reiner Publishing.

Scott, J. 1971. *International of Norms*. Prentice-Hall.

Selten, R. and R. Stoecker. 1986. "End Behavior in Sequences of Finite Prisoner's Dilemma Supergames: A Learning Theory Approach." *Journal of Economic Behavior and Organization* **7**:47–70.

Sherif, M. 1938. *The Psychology of Social Norms*. Harper and Brothers.

Simon, H. 1982. *Models of Bounded Rationality: Behavioral Economics and Business Organization*. MIT Press.

Skyrms, B. 1986. "Deliberational Equilibria." *Topoi* **5**:59–67.

—— 1989. "Deliberational Dynamics and the Foundations of Bayesian Game Theory." In J. E. Tomberlin, ed., *Epistemology, Philosophical Perspectives*. Vol. 2. Ridgeview: Northridge.

—— 1990. *The Dynamics of Rational Deliberation*. Harvard University Press.

Sober, E. 1992a. "Stable Cooperation in Iterated Prisoners' Dilemmas." *Economics and Philosophy* **8**:127–39.

—— 1992b. "The Evolution of Altruism: Correlation, Cost and Benefits." *Biology and Philosophy* **7**:177–87.

Stroll, A. 1987. "Norms." *Dialectica* **41**:7–22.

Sugden, R. 1986. *Economics of Rights, Cooperation and Welfare*. Oxford: Blackwell.

Taylor, P. and L. Jonker, 1978. "Evolutionarily Stable Strategies and Game Dynamics." *Mathematical Biosciences* **40**:145–56.

Terhune, K. 1968. "Matrices, Situations, and Interpersonal Conflict within the Prisoner's Dilemma." *Journal of Personality and Social Psychology* **3**:1–24.

Thibaut, T. and H. Kelley. 1959. *The Social Psychology of Groups*. Wiley.

Trivers, R. 1971. "The Evolution of Reciprocal Altruism." *Quarterly Review of Biology* **46**:35–57.

—— 1985. *Social Evolution*. Cummings.

Ullmann-Margalit, E. 1977. *The Emergence of Norms*. Oxford: Oxford University Press.

Winson, P. 1980. "Learning and Reasoning by Analogy." *Communications of the ACM* **23**:689–703.

Young, P. 1993. "The Evolution of Conventions." *Econometrica* **61**:57–84.

—— and D. Foster. 1991. "Cooperation in the Short and in the Long Run." *Games and Economic Behavior* **3**:145–56.

3

On the dynamics of social norms

PIER LUIGI SACCO

Abstract

This paper proposes a model for the study of social norms and of their dynamics in a repeated prisoner's dilemma context. The distribution of players' actions across the population is assumed to evolve according to the replicator dynamics. A behavioral rationale for this specification is provided. Some results on the relative efficiency of cooperative vs. noncooperative norms are also provided, as well as a discussion of the relative performance of various constitutional metanorms that may be used to revise relatively inefficient norms at predetermined dates.

[In the Arnstein and Feigenbaum (1967) experiment] persons of different religious persuasions were asked to play a game of the prisoner's dilemma variety; in this game, noncooperative behavior improved considerably the lot of the noncooperative player provided the other player's behavior remained cooperative. Conversely, cooperative players fared quite poorly if the other players were noncooperative. It turned out that the Quakers, as might be expected, ranked quite high in terms of the truthfulness and cooperation of their responses, but low in terms of their economic rationality. This result is curious because in real life, Quakers are usually considered one of the wealthiest minority groups in the United States (Allport, 1958, p. 72). (Akerlof [1983], p. 55)

1. INTRODUCTION

The issue of the rationality of cooperative behavior is a long-standing one in economically motivated game theory. The classical game-theoretic setting for the analysis of cooperative behavior, namely, the prisoner's dilemma game, shows that cooperation is in fact a strictly dominated strategy if the game is repeated only a finite number of times; for the infinitely repeated version of the game, the standard folk theorem states that for a large enough discount factor the

Pier Luigi Sacco, Department of Economics, University of Florence, Via Montebello, 7, 50123 Firenze, Italy. E-mail: sacco@ifistat.bitnet.

equilibrium pattern of cooperation/noncooperation is basically indeterminate on aprioristic grounds.

Many different ways of escape from this impasse have been tried; one of them borrows from biologically motivated game theory. In this respect, the solution concept known as evolutionary stability, introduced by Maynard Smith (see, e.g., Maynard Smith [1982]), in its many variations (see, among others, Crawford [1991]; Samuelson [1991]; Swinkels [1992]), has attracted considerable interest as an evolutionarily motivated refinement of Nash equilibrium (see, e.g., van Damme [1991]). Fudenberg and Maskin (1990) have in fact proved that under certain conditions (the most important being finite complexity of strategies and a moderately noisy environment), evolutionary stability selects "relatively cooperative" outcomes (i.e., outcomes that entail at least a certain amount of cooperation) in the inifinitely repeated prisoner's dilemma. Binmore and Samuelson (1992) prove an analogous, somewhat stronger result in a different context.

The evolutionary approach to cooperation has also tried to embed players' strategic interaction in a richer framework that allows for the existence of social institutions, in the spirit of Granovetter (1985) and, more to the point, Axelrod (1986). For example, Sugden (1989) has proposed to characterize social conventions as evolutionarily stable states of suitable games; Bicchieri (1990) goes beyond providing a dynamic evolutionary rationale for social norms. Okuno-Fujiwara and Postlewaite (1991) and Kandori (1992) build an explicit dynamic model of rational cooperation based on a given environment of social norms when players have a limited knowledge of the state of the society (i.e., of the distribution of behavioral types across the population). Gilboa and Matsui (1991) study a best-response dynamics leading to socially stable outcomes that need not be Nash (see also Matsui [1992]).

Among the many possible dynamic specifications of evolutionary processes, the most studied is perhaps the so-called replicator dynamics, again a biologically motivated notion; it postulates that strategies (behaviors) that have higher than average fitness tend to be selected at the expense of others. If fitness is reinterpreted in terms of payoffs (either literally meant as the number of players' offspring or, more broadly, as money payments), these dynamics have a natural game-theoretic interpretation.[1] Blad (1986) provides a pioneering analysis of the replicator dynamics generated by the prisoner's dilemma game in a random-matching environment. Friedman (1991) and Nachbar (1990) provide general treatments of wider classes of selection dynamics (that include the replicator dynamics as a special case) and analyze their relationships with traditional game-theoretic solution concepts. Further results on the game-theoretic characterization of limit outcomes of selection dynamics (either in continuous or in discrete time) are found in Samuelson and Zhang (1992), Dekel and Scotchmer (1992), and Cabrales and Sobel (1992).

In this paper I analyze the dynamics of cooperation in the prisoner's dilemma game with random matching in continuous time within an environment of social norms, which are not fixed as in Okuno-Fujiwara and Postlewaite (1991) and Kandori (1992) but evolve themselves through time. Unlike the previous authors, though, I assume that the current state of the society is directly observed by all players and that the enforcement of the social norm is not carried out by players but by an outside referee. I adopt the replicator-dynamics specification, which is, however, reinterpreted in order to avoid the unnaturalities that arise when it is literally translated from a biological to a socioeconomic setting. More specifically, I assume that players' payoffs are in fact money payments rather than offspring and that selection operates at the *behavioral* level in the sense that relatively more rewarding strategies tend to be *imitated* by an increasing number of players at the expense of others. In other words, I take the replicator dynamics as a model of the selection among behaviors rather than among individuals: successful behaviors spread over the population of players, whereas unsuccessful behaviors die out. Since in my interpretation experience can be "inherited" through imitation by later "generations" of behaviors, the evolutionary dynamics I study are somewhat closer to the Lamarckian rather than to the Darwinian model of selection. A more detailed discussion of these issues may be found in Sacco (1994).

In this paper I show that a society that manages to enforce cooperation through a certain set of social norms may become wealthier than less cooperative societies. Indeed, the quotation at the top of the paper (while hinting at this possibility) seems to suggest that, when interacting with players from less cooperative societies, players who grew up in a cooperative environment tend to burn their fingers (and nevertheless persist in their behavior, due to the powers of the socialization process to which they have been exposed). This paradox is largely an artifact of the extreme, special conditions under which the experimental game is played. It is hard, to say the least, to evaluate the rationality of cooperative behavior when the environmental conditions are so markedly different from those that led to its "selection," that is, to the emergence of a social convention that makes of cooperation a "culturally obvious" mode of play. Once this is realized, it is possible to build sets of norms (viz., societies) that actually enforce cooperation in a way that is robust to the invasion of defective players. Moreover, the social outcome brought about by such norms may be more efficient than that of less cooperative societies. When this is the case, the issue becomes how a noncooperative society can change its set of social norms in order to promote cooperation.

The structure of the rest of the paper is the following. In section 2 the basic model is presented. Section 3 studies the dynamics of the model. Section 4 investigates the relationships between social norms and (asymptotic) social efficiency. Section 5 compares the (asymptotic) efficiency of various social environments. Section 6 studies the dynamics of social norms under various adjustment rules. Section 7 describes possible extensions of the model.

2. THE MODEL SOCIETY

I consider a society made up of a continuum of individuals who are randomly matched to play the following version of the prisoner's dilemma game:

$$
\begin{array}{ccc}
 & C & D \\
(1) \quad C & (\beta, \beta) & (-\alpha, \alpha) \\
D & (\alpha, -\alpha) & (0, 0)
\end{array}
$$

where clearly $\alpha > \beta > 0$. This specification, which is slightly different from the usual one, has been chosen to increase players' incentive to defect. Some of my results will therefore apply a fortiori to more conventional specifications.

In order to qualify as a society, my community of players must act on the basis of some set of social norms. I therefore assume that there is a referee who observes each round of play and decides to give players a reward whose level depends on their actions; negative rewards (i.e., punishments) are allowed. More specifically, if a player played cooperatively, she receives the amount $\gamma > 0$ for sure; otherwise, the player receives γ with probability θ and $-\gamma$ with probability $1 - \theta$. In other words, the parameter θ measures the societal propensity to punish noncooperative behavior; a high θ means that the society has only a moderate concern with the enforcement of cooperation, and vice versa. (For the sake of simplicity, I imagine that the resources paid by defectors when they are punished are burnt or kept idle by the social authority.) In symbols,

$$
(2) \qquad\qquad p(\gamma|C) = 1,
$$

$$
(3) \qquad\qquad p(\gamma|D) = \theta, \; p(-\gamma|D) = 1 - \theta.
$$

Time is continuous. At time t, a proportion μ of the population intends to play C if called to play, whereas a proportion $1 - \mu$ intends to play D. In other words, μ is the proportion of cooperators at time t. μ is perfectly observable; players do not know, however, the behavioral type of the individual player with which they are matched. Given the structure of the game, players look at ex ante (i.e., expected) payoffs. The expected payoff from playing cooperatively is thus given by

$$(4) \qquad E\pi_C = \mu\beta + (1-\mu)(-\alpha) + \gamma = (\alpha+\beta)\mu + \gamma - \alpha.$$

Accordingly, the expected payoff from playing noncooperatively is

$$(5) \qquad E\pi_D = \mu\alpha + (1-\mu)0 + \theta\gamma + (1-\theta)(-\gamma)$$
$$= \alpha\mu + \gamma(2\theta - 1).$$

The society's (expected) wealth given the distribution of types across the population is thus

$$(6) \qquad W(\theta, \mu) = \mu E\pi_C + (1-\mu)E\pi_D = \beta\mu^2 + 2\gamma(1-\theta)\mu + \gamma(2\theta - 1).$$

Two interesting conclusions can be drawn already; in the first place, we have

$$(7) \qquad \frac{\partial W(\theta, \mu)}{\partial \theta} = 2\gamma(1-\mu) > 0$$

for μ less than one. That is, the less likely the punishment, the higher (ceteris paribus) aggregate wealth. On the other hand,

$$(8) \qquad \frac{\partial W(\theta, \mu)}{\partial \mu} = 2[\beta\mu + \gamma(1-\theta)] > 0.$$

In other words, aggregate wealth increases with the proportion of cooperators. As to the difference between (expected) payoffs, one has

$$(9) \qquad D(\theta, \mu) = E\pi_C - E\pi_D = \beta\mu - \alpha + 2\gamma(1-\theta).$$

It thus follows that

$$(10) \qquad \frac{\partial D(\theta, \mu)}{\partial \theta} = -2\gamma < 0,$$

$$(11) \qquad \frac{\partial D(\theta, \mu)}{\partial \mu} = \beta < 0.$$

These results are intuitively plausible; if the probability of punishment decreases, cooperative behavior performs relatively worse, whereas if the proportion of cooperators increases, cooperative behavior performs relatively better (ceteris paribus).

In conclusion, an increase in θ tends to raise (expected) wealth *given* μ but, on the other hand, it reduces the (expected) difference between payoffs making defection relatively more attractive (or at least less unattractive); if, as a consequence of this, μ decreases, a decrease of (expected) aggregate wealth follows. The net effect is therefore ambiguous: from the point of view of the social authority, one would like to bring about cooperation while not punishing defection too often.

3. Evolutionary Dynamics

I have so far said nothing as to how the social norm θ and the proportion of cooperators μ are determined and change through time. I assume that the initial values of the two parameters are independently and uniformly distributed over the unit square. In this section I assume, moreover, that θ stays fixed at its initial value. I will remove this assumption later on to see how social norms evolve themselves in the long run.

As anticipated, I model the evolutionary dynamics of cooperation via the so-called replicator dynamics:

$$(12) \qquad \dot{\mu} = \mu(1 - \mu)D(\mu, \theta) = \mu(1 - \mu)[\beta\mu + 2(1 - \theta)\gamma - \alpha].$$

It is easy to check that (12) dictates that the proportion of cooperators increases if and only if the payoff of cooperation is larger than the average population payoff; since players have to choose between two options only, this amounts to requiring that the payoff of cooperation be larger than the payoff of defection. The dynamics (12) may be interpreted as a model of a population of players with an asynchronous timing of decisions according to which only a measure-zero subset of players decides whether to change its type (cooperator or defector) at each given time (so to induce a smooth, continuous time aggregate dynamics as discussed by Gray and Turnovsky [1979]); more specifically, players who are called to revise their decision are willing to change their type if and only if, at the time the decision is made, their actual strategy does worse than the alternative one.

Notice that $\mu = 0$ and $\mu = 1$ are always stationary points of the dynamics; that is, pure populations, either of cooperators or of defectors, self-reproduce if nothing else happens. There could also be stationary points in the interior of the unit interval. If existing, these points $\hat{\mu}$ are such that $D(\hat{\mu}, \theta) = 0$; from (9) one therefore has

$$(13) \qquad \hat{\mu} = \frac{\alpha - 2(1 - \theta)\gamma}{\beta}.$$

$\hat{\mu}$ actually exists when it belongs to the interior of the unit interval, that is, when

$$(14) \qquad 1 - \frac{\alpha}{2\gamma} < \theta < 1 - \frac{\alpha}{2\gamma} + \frac{\beta}{2\gamma}.$$

Since players choose between two options only, it is easy to prove that a given distribution of behavioral types across the population is evolutionarily stable if and only if it is asymptotically stable under (12) (see, e.g., Hofbauer and Sigmund [1988]). (Clearly, stable stationary points of [12] are a fortiori also Nash equilibria.)

In order to characterize the dynamic behavior of (12) it is useful to distinguish several cases. Let me introduce some useful terminology first. We say that "cooperation dominates" if $\mu = 1$ is globally stable under (12), that is, if $\mu \to 1$ for all initial conditions $\mu_0 \neq 0$. Analogously, we say that "defection dominates" if $\mu = 0$ is globally stable under (12), that is, if $\mu \to 0$ for all $\mu_0 \neq 1$. Finally, we say that "bistability" occurs if $\hat{\mu}$ lies in the interior of the unit interval and $\mu \to 1$ for $\mu_0 > \hat{\mu}$, $\mu \to 0$ for $\mu_0 < \hat{\mu}$. We then have[2]

Proposition 1 *Let* $\alpha < 2\gamma$. *Then for* $0 \leq \theta \leq 1 - \alpha/2\gamma$ *cooperation dominates. For* $1 - \alpha/2\gamma < \theta < 1 - \alpha/2\gamma + \beta/2\gamma$ *bistability occurs. Finally, for* $1 - \alpha/2\gamma + \beta/2\gamma \leq \theta \leq 1$ *defection dominates.*

Proposition 1 can be interpreted as follows; when $\alpha < 2\gamma$ the temptation to defect faced by players is moderate in relative terms, that is, with respect to the size γ of the punishment.[3] Consequently, if defective behavior is punished often enough, cooperation is enforced. If defection is punished less often, cooperation is enforced only if the initial proportion of cooperators is large enough. Of course, the stronger[4] the social norm, the larger the basin of attraction of $\mu = 1$ (i.e., the less binding the "high enough" constraint). Finally, if defection is seldom punished (or, in the limit, never punished), defection spreads over.

The following results suggest that as the temptation to defect becomes more substantial, there is less room for cooperation however strong the social norm.

Proposition 2 *Let* $2\gamma \leq \alpha < 2\gamma + \beta$. *Then if* $0 \leq \theta < 1 - \alpha/2\gamma + \beta/2\gamma$ *bistability occurs. If instead* $\theta \geq 1 - \alpha/2\gamma + \beta/2\gamma$ *defection dominates.*

Proposition 3 *Let* $\alpha \geq 2\gamma + \beta$. *Then defection dominates for every* $0 \leq \theta \leq 1$.

4. SOCIAL NORMS AND ASYMPTOTIC SOCIAL WEALTH

In this section I explore how asymptotic social wealth varies with the social norm adopted. I consider the case $\alpha < 2\gamma$. The other cases are treated in an analogous way and are left to the interested reader.

Assume that defection is often punished, that is, $\theta < 1 - \alpha/2\gamma$. In this case, $\mu^* = 1$ (I denote stable stationary points with starred superscripts). To measure the strength of the punishment, I adopt the parametrization $0 \leq \varepsilon_C \leq 1 - \alpha/2\gamma$ and write

$$(15) \qquad \qquad \theta \equiv 1 - \frac{\alpha}{2\gamma} - \varepsilon_C.$$

That is, $\varepsilon_C = 1 - \alpha/2\gamma$ denotes the strongest possible social norm (punishment for defectors is certain), whereas $\varepsilon_C = 0$ corresponds to the "threshold" case (i.e., the weaker social norm that is compatible with the global enforcement of cooperation). It is easy to check that, in this case,

$$(16) \qquad W^*(\theta) \equiv \tilde{W}^*(\varepsilon_C) = \beta + \gamma.$$

We thus have

Fact 1 *In a society in which cooperation dominates, social wealth is asymptotically independent of the social norm adopted.*

Fact 1 points out that if cooperation dominates, no player defects eventually and thus punishment is never called for; as a consequence, the strength of the social norm cannot affect social wealth at the equilibrium.

I now move to the case where θ lies between $1 - \alpha/2\gamma$ and $1 - \alpha/2\gamma + \beta/2\gamma$. Within this range of values of θ, bistability occurs; I index it via the parametrization $0 < \varepsilon_B < 1$ and write

$$(17) \qquad \theta \equiv 1 - \frac{\alpha}{2\gamma} + \frac{\beta}{2\gamma}(1 - \varepsilon_B).$$

More specifically, $\varepsilon_B = 0$ denotes the threshold norm between bistability and defection dominance, whereas $\varepsilon_B = 1$ denotes the threshold norm between bistability and cooperation dominance.

Let us start from the subcase $\mu_0 < \hat{\mu}$, that is, $\mu^* = 0$ (the initial proportion of cooperators is not large enough to enforce cooperation). One then has

$$(18) \qquad W^*(\theta) = \gamma(2\theta - 1),$$

from which it follows that

$$(19) \qquad \tilde{W}^*(\varepsilon_B) = \gamma - \alpha + \beta(1 - \varepsilon_B).$$

As could be expected, asymptotic social wealth decreases as ε_B (i.e., the probability of punishment) grows, *given* that the social norm adopted causes bistability.

On the other hand, if $\mu_0 > \hat{\mu}$, $\mu^* = 1$ and, consequently,

$$W^*(\theta) = \tilde{W}^*(\varepsilon_B) = \beta + \gamma.$$

Finally, in the case $1 - \alpha/2\gamma + \beta/2\gamma \leq \theta \leq 1$, the convenient parametrization is $0 \leq \varepsilon_D \leq (\alpha - \beta)/2\gamma$; one therefore writes

$$(20) \qquad \theta \equiv 1 - \frac{\alpha}{2\gamma} + \frac{\beta}{2\gamma} + \varepsilon_D,$$

where $\varepsilon_D = (\alpha - \beta)/2\gamma$ denotes the case in which defectors are never punished. This yields

$$W^*(\theta) = \gamma(2\theta - 1),$$

viz.

$$(21) \qquad \tilde{W}^*(\varepsilon_D) = \gamma - \alpha + \beta + 2\gamma\varepsilon_D.$$

Once again, asymptotic social wealth decreases as the probability of punishment grows, ceteris paribus.

<h2 style="text-align:center">5. Intersocietal Comparisons of Wealth</h2>

In this section I compare the relative performance of various social environments. Once again I focus my attention on the case $\alpha < 2\gamma$, leaving the analogous treatment of the remaining ones to the interested reader.

Let us compare first the asymptotic wealth of a society in which cooperation dominates with that of a society in which defection does. From equations (16) and (21), it follows that the latter kind of society is richer when

$$(22) \qquad \varepsilon_D > \frac{\alpha}{2\gamma}.$$

In view of the restrictions on ε_D (see section 4 above), this condition is never met. We thus have

Fact 2 *Let $\alpha < 2\gamma$. Then a society in which cooperation dominates is always asymptotically richer than a society in which defection dominates.*

This is an interesting result. When $\alpha < 2\gamma$, even in those societies in which it is individually rational to defect in spite of the possibility of punishment, defection is inefficient at the aggregate level; that is, the society would benefit by strengthening the social norm enough to make cooperation enforceable. I return to this point in section 6.

We can, moreover, compare the performance of a society in which defection dominates with that of a "bistable" society for which $\mu_0 < \hat{\mu}$. The former kind of society is richer if

$$(23) \qquad \varepsilon_D > -\frac{\beta}{2\gamma}\varepsilon_B,$$

a condition that is always met. Thus,

Fact 3 *A society in which defection dominates is always asymptotically richer than a "bistable" society in which the initial proportion of cooperators is too low to make cooperation enforceable.*

Fact 3 is easily rationalized. A bistable society requires a stronger norm than a society in which defection dominates. Nevertheless, in both cases all players defect at the equilibrium. For this reason, social wealth is increased if the probability of punishment is reduced (at best, set to zero: the social authority surrenders to the status quo).

By the same token one can show that

Fact 4 *A "bistable" society in which the initial proportion of cooperators is high enough to enforce cooperation is always asymptotically richer than a society in which defection dominates (and, a fortiori, of a "bistable" society in which the initial proportion of cooperators is too low to enforce cooperation). Actually, it is as asymptotically rich as a society in which cooperation dominates.*

These results depend once again on the fact that whenever all players cooperate at the equilibrium, the strength of the social norm is irrelevant. This of course does not mean that the social norm is also irrelevant during the *transition* toward equilibrium. For example, in a bistable society with a high enough initial proportion of cooperators, cooperation is enforced at a lower cost than that paid by a society in which cooperation dominates, because in the latter defectors are punished more often.

One can also make comparisons between societies characterized by different payoff matrices of the base game. These payoff matrices are beyond the control of the social authorities; such comparisons have therefore a different nature than the previous ones and concern, at least in part, intersocietal differences that are not amenable to social reform. One comparison has special interest: that between a society in which cooperation dominates and $\alpha < 2\gamma$ and a society in which defection dominates and $\alpha > 2\gamma$.[5] I am thus comparing a society in which players are only moderately tempted by defection and cooperation is enforced by the social norm with a society in which the temptation to defect is more substantial and players actually defect.

Let α' be the payoff of defection (against a cooperator) in the latter kind of society; we can write $\alpha' = 2\gamma + \delta$, where $\delta > 0$. If defection dominates, however, no player has a chance to meet a cooperator at the equilibrium. As a consequence, the actual value of δ is irrelevant at the equilibrium, and the noncooperative society's asymptotic wealth is at most equal to γ (this occurs when defectors are never punished). Therefore, the cooperative society is always more efficient, since $\beta > 0$. Thus,

Fact 5 *A society in which cooperation dominates and $\alpha < 2\gamma$ is always asymp­totically richer than a society in which defection dominates and $\alpha > 2\gamma$.*

(In view of Fact 5, in Fact 2 the restrictive condition $\alpha < 2\gamma$ is no longer needed provided that α is the only payoff parameter that is varied from one society to the other.)

6. THE DYNAMICS OF SOCIAL NORMS

In order to study the dynamics of social norms in a given environment, I now consider two neighboring societies that are both able to observe the social outcome that occurs in the other. Call these societies A and B. I assume that both societies are endowed with some sort of "constitutional" metanorm, that is, a rule that specifies when and how the social norm in use must be changed. More specifically, I postulate that the constitutional metanorms of both societies prescribe that the respective social norms may be changed at predetermined dates, called constitutional dates. I assume, moreover, that the order of magnitude of the time needed to reach the population equilibrium starting from the initial distribution of behavioral types is substantially smaller than the order of magnitude of the time between two constitutional dates and that the constitutional dates of the two societies are simultaneous. Consequently, when a constitutional date comes, both societies are at a population equilibrium.[6]

Given the above assumptions, a simple and natural choice of the criterion followed in the constitutional revision of social norms is the comparison of the society's aggregate outcome with that of the neighboring society, that is, the comparison of the relative levels of social wealth. If, at a constitutional date, society A finds that its level of social wealth is lower than that of society B, it concludes that the social norm adopted is inefficient and chooses a new one. Society A may then merely choose to "imitate" its more efficient neighbor, selecting a new norm that is "closer" to that of society B, or may adopt more sophisticated revision rules. In the case of straightforward imitation, we speak of adaptive constitutional reform.

Furthermore, I assume that, after each constitutional date, a small fraction of players dies and is substituted by an equal number of newly born players whose (initial) type (viz., cooperator or defector) is stochastic; this amounts to perturbing slightly the equilibrium distribution of types. If, therefore, the constitutional reform is large enough to bring the inefficient society (say society A) into a new dynamic regime, a new population equilibrium will be reached before the next constitutional date comes.

Formally, I postulate that the actual change of the social norm undertaken by the inefficient society (say society A) at the constitutional date is proportional to the observed efficiency gap:

$$(24) \qquad \Delta\theta_A = \eta\Delta W^*(\theta_A, \theta_B),$$

where $\Delta\theta_A = \theta'_A - \theta_A$ (θ'_A being the new social norm adopted by society A), $\Delta W^*(\theta_A, \theta_B) = W^*(\theta_A) - W^*(\theta_B) < 0$ and where the sign of η is determined by the type of revision rule used. In the case of an adaptive constitutional reform, $\eta > 0$ if the more efficient society is cooperative, whereas $\eta < 0$ if it is noncooperative. A relatively small $|\eta|$ implies a relatively "conservative" society, whereas a relatively large $|\eta|$ is evidence of a society more inclined to change.

Clearly, θ'_A is subject to the constraints

(25) $$\theta'_A = \min(1, \theta_A + \eta \Delta W^*(\theta_A, \theta_B)) \quad \text{if} \quad \eta < 0;$$

(26) $$\theta'_A = \max(0, \theta_A + \eta \Delta W^*(\theta_A, \theta_B)) \quad \text{if} \quad \eta > 0.$$

Of course, not all possible constitutional reforms give the desired results; basically, the success of a constitutional reform in a given situation depends on the sign and value of η. For example, assume that society A, the inefficient one, is noncooperative, whereas society B, the efficient one, is cooperative (as above, I assume that $\alpha < 2\gamma$ and that the payoff matrix specification is the same for both societies). If following an adaptive revision rule, when the constitutional date comes, society A deliberates to punish defection more often than in the past. In this case,

$$\Delta W^*(\theta_A, \theta_B) = 2\gamma \varepsilon_D - \alpha.$$

Since the perturbation of the equilibrium distribution of types caused by the deaths and births that occur after the constitutional date is small and since $\mu_A^* = 0$, μ_0 in the next ordinary phase will be close to zero. Therefore, in order to enforce cooperation, the constitutional reform must be large enough to make cooperation dominate;[7] a moderate change that brings θ into the bistable range will not manage in general to put the society on the way to the cooperative population equilibrium.

The size of the punishment γ could in principle be subject to constitutional revision as well. In this paper, however, I have chosen to assume γ fixed for a given society, just like the payoffs of the base game (1). In fact, a society could always manage to enforce cooperation if it is tough enough against defectors: if, for example, the punishment is made larger than the maximum payoff that can be attained in the base game (namely, α), cooperation may become a dominant choice even if punishment does not occur with certainty. But this policy is not really feasible for the social authority, because players would refuse to endorse punishment schemes they find "unfairly" tough. Clearly, different societies have different perceptions of how tough a punishment scheme must be to be judged unfair; such differences in perception can be represented in terms of differences in the α/γ ratio: the smaller this ratio, the more "authoritarian" the society, that is, the more willing it is to support tough punishment of defectors. To fix ideas, in this paper I assume that, for every given society, γ is set at the highest "feasible" level. Raising the punishment is therefore not possible for the social authority. On the other hand, I also rule out the possibility of *lowering* the punishment. By lowering γ, a society redefines its attitude toward defection, that is, it questions the norm itself, whereas an increase of θ (that at first sight should generate analogous effects) simply amounts to adjusting the level of enforcement. This point should not be underplayed: as is apparent from the analysis of section 3, changes in γ may cause changes in the menu of possible regimes for the social dynamics; changes in θ can only cause less dramatic effects, that is, regime change within a given menu. This important difference also reflects into intersocietal comparisons of wealth, as shown in section 5: the critical parameter that determines the relative wealth of cooperative vs. noncooperative societies is α/γ. When conceived in its fully generality, institutional change is, therefore, likely to have a hierarchic structure: if the enforcement of cooperation becomes an issue for the society, the first point on the constitutional agenda will be the enforcement of the norm (i.e., an adjustment of θ); questioning the norm itself (i.e., changing γ if feasible) is a last resort. A sensible theoretical treatment of this more general case in which both θ and γ may vary then requires more careful and detailed modeling of the social and political organization of the community of players, of the decision-making problem faced by the social authority, and, more fundamentally, modeling the birth and emergence of norms *and of their enforcement mechanisms* (see, e.g., Kliemt [1990]).

This said, it is easy to check that the more authoritarian a society, that is, the smaller α/γ, the easier it is ceteris paribus to reach the cooperative equilibrium. The actual constitutional reform needed to turn the noncooperative, inefficient society A into a cooperative, efficient one is such that

$$(27) \qquad 1 - \frac{\alpha}{2\gamma} + \frac{\beta}{2\gamma} + \varepsilon_D + \eta(2\gamma\varepsilon_D - \alpha) \leq 1 - \frac{\alpha}{2\gamma}.$$

(Clearly, in this case $\eta > 0$.) The critical value of η beyond which the constitutional reform is successful is therefore

$$(28) \qquad \eta^* = \frac{\beta + 2\gamma\varepsilon_D}{2\gamma(\alpha - 2\gamma\varepsilon_D)}.$$

Notice how the minimal constitutional reform needed to reach efficiency depends on "how far" the original social norm was from the range of values of θ for which cooperation dominates. More specifically, the larger ε_D (i.e., the less often defection was punished in the inefficient society), the larger the constitutional reform needed to enforce cooperation. This result suggests that relatively conservative societies that are not inclined to big constitutional reforms could lock themselves into inefficient social outcomes in the long run.

Assume for example that (28) is violated, that is, society A follows an adaptive revision rule but is nevertheless too conservative to choose a new norm that allows the enforcement of cooperation. Many possible scenarios may then be considered, some of which are of particular interest.

Scenario 1. Straightforward adaptive reform

If the actual η is smaller than η^* as defined in (28), the constitutional reform does not manage to turn the inefficient, noncooperative society into an efficient, cooperative one. As a consequence, players still defect as the equilibrium and asymptotic social wealth actually *decreases* with respect to the previous social norm. If, in spite of this, the society sticks to the adaptive constitutional metanorm – it keeps on punishing defection more and more at each subsequent constitutional date – cooperation will eventually be enforced. Note in fact that, since after each unsuccessful constitutional reform the aggregate wealth of the inefficient society decreases, the efficiency gap thereby increases. Being the size of the adjustment is proportional to the efficiency gap, even relatively conservative societies will undertake bigger and bigger constitutional reforms after each failure. As a consequence, the eventual enforcement of cooperation will be accelerated. In conclusion, a straightforward adaptive constitutional metanorm of this sort eventually leads the society to the efficient outcome.

Scenario 2. Conditioned reform

The previous scenario sounds somewhat unrealistic, because it requires that the inefficient society persist in the imitation of the neighboring society in spite of the past record of failures, however long. This seems unlikely if the feasibility of a constitutional reform depends at least to some degree on social consensus. A more sophisticated society could then adopt a less straightforward constitutional metanorm that links later constitutional deliberations to the current reform's actual performance. For example, if the constitutional reform actually produces a *decrease* in social wealth, the society might deliberate that punishing defectors more often was not a good idea after all and that, say, defection is an unavoidable, socially idiosyncratic mode of play that cannot be wiped out. As a consequence, at the next constitutional date the inefficient society would switch back to a norm that punishes defection less often, locking itself into an inefficient institutional setting. An analogous story could be told for a less impatient society that deliberates to switch back after k unsuccessful constitutional reforms. Of course, the larger k the more likely that the society manages to enforce cooperation in less than k steps *given* η. This pattern of adjustment might be rationalized by bringing political institutions into the picture: if the incumbent party X bets on the constitutional reform that turns out to be unsuccessful, it loses the next political elections at the advantage of party Y, whose political platform is based on the restoration of the previous status quo. We thus have[8]

Fact 6 *An inefficient society which is relatively conservative can be sure to enforce cooperation only if it follows the straightforward adaptive constitutional metanorm; more sophisticated metanorms may lead to persistent inefficiency.*

Scenarios 3–4. Adaptive vs. sophisticated reform "far from efficiency"

A different situation arises when both societies are noncooperative. In this case, the adaptive metanorm dictates that the relatively inefficient society, the one that punishes defection relatively more often without managing to enforce cooperation, adopts a weaker social norm (i.e., increases θ). This reform actually increases asymptotic social wealth but (at least as far as $\alpha < 2\gamma$) is globally inefficient. In such a case, more sophisticated revision rules would be useful but could not be justified on adaptive grounds. For the reasons explained above (e.g., political factors), a sophisticated search rule is once again harder to enforce the more conservative the society (i.e., the larger the number of constitutional reforms needed to find the efficient social norm).

Fact 7 *When both societies are noncooperative, the inefficient one may enforce cooperation only if it is not too conservative and if it adopts a sophisticated metanorm that is not adaptively motivated (actually, a metanorm that allows for reforms that go against the available intersocietal evidence).*

No general case for adaptive vs. sophisticated metanorms may thus be made. If one of the societies has already reached the efficient social outcome, straightforward imitation is better than sophisticated search. If, however, both societies are far from efficiency, more sophisticated search rules are needed.

I now consider more intriguing possibilities.

Scenario 5. Hopeless reform

Assume that payoff specifications are different for the two societies (in the way described at the end of section 5) and that, in particular, society A is noncooperative such that $\alpha > 2\gamma$, whereas society B is cooperative such that $\alpha < 2\gamma$. In this case, straightforward imitation of the efficient society is simply hopeless, because the payoff matrix specification for society A does not allow for a cooperative population equilibrium (at least if the initial proportion of cooperators is low). This is actually the case where the efficiency gap between the two societies is indeed caused by factors that are beyond the control of the social authorities.[9] Adoption of the straightforward adaptive metanorm by the inefficient society would therefore produce very bad outcomes now. In this case, the better search rule is to go *against* the available evidence from the other society (i.e., to adopt norms that punish defection less often), surrendering to the noncooperative status quo in spite of the encouraging example of the neighboring community.

Fact 8 *When there are a noncooperative society such that $\alpha > 2\gamma$ and a cooperative society such that $\alpha < 2\gamma$, the former should never use an adaptive metanorm. Once again, a sophisticated metanorm that is not adaptively motivated is needed to reach the best feasible social outcome.*

As a final possibility, I consider the case in which both societies are bistable but A is actually noncooperative, whereas B is cooperative. In this case one can have the apparently paradoxical situation in which both societies use the same social norm (the same θ) but one is more efficient than the other. The differences between the two societies are here basically a product of historical rather than of social forces (i.e., of the difference between the initial realizations of μ_0). Clearly, no obvious search rule is eligible now for the inefficient society, which is likely to trace its relative inefficiency back to intrinsic "cultural" differences. A couple of related interesting scenarios may, however, be examined.

Scenario 6. Overshooting

Assume that both societies are bistable, A is inefficient, and $\theta_A > \theta_B$. In this case, if A is not too conservative, an adaptive constitutional reform could "overshoot" bistability, putting society A within the efficient range of θ where cooperation dominates. Thus, in spite of the small initial proportion of cooperators, cooperation is eventually enforced and society A reaches the efficient outcome.

Scenario 7. Perverse imitation

A perverse case may occur when both societies are bistable, A is noncooperative, B is cooperative, and $\theta_A < \theta_B$. If in this instance society A uses the adaptive metanorm, at the constitutional date it deliberates to *increase* θ (i.e., to punish defection less often), thus increasing asymptotic social wealth but getting further away from the efficient range of θ. Once again, in order to reach an efficient outcome, society A must adopt a metanorm that is not adaptively motivated.

Fact 9 *If both societies are bistable, straightforward imitation may yield the efficient outcome only if $\theta_A > \theta_B$ (A being the inefficient society). Otherwise, a sophisticated, not adaptively motivated, search rule is needed.*

7. EXTENSIONS

Many possible extensions of the simple framework used in this paper may of course be devised. One interesting possibility would be that of considering punishment schemes that depend not only on the actual play but also, to some extent, on players' past history of play. In this case, even in societies in which cooperation dominates, asymptotic social wealth would depend on the social norm adopted, thus leaving room for the possibility of noncooperative societies being more efficient than cooperative ones under certain conditions. More specifically, the relative efficiency of cooperation vs. defection would depend on the values of both ε_C and ε_D. Similar possibilities emerge, assuming that the monitoring device used by the referee to check players' actions is noisy, that is, cooperators may be punished by mistake, and the frequency of mistakes depends on the frequency with which defectors are punished, that is, on the strength of the social norm. In this case, cooperative societies need not be efficient, because of adverse noise effects.

I have assumed also that resources paid by defectors who are punished are burnt or kept idle by the social authority. If these resources can be invested and if investments are rewarding enough, a noncooperative society in which the social authority may draw on a large resource pool to finance socially useful projects might turn out to be more efficient than a cooperative society in which few or no funds for such projects become available.

Further, one could explore more closely the various possible kinds of sophisticated search rules (metanorms), their relative performance, and their robustness to various environmental conditions. Alternatively, one could deal explicitly with the political process that lies behind constitutional reforms, that is, one could formally model political parties that compete on the basis of different platforms of reforms. The relationships between social efficiency and political factors could then be investigated.

Finally, a more detailed and structured modeling of social norms could be attempted. The discussion of Elster (1989), as well as the issues he raises, might be very useful in this respect.

I leave the exploration of various combinations of these and other possible extensions for future research.

ACKNOWLEDGMENT

I thank the audience at the Second Workshop on Knowledge, Belief and Strategic Interaction, Castiglioncello, June 18–22, 1992, and in particular Robert Axelrod, Andrea Battinelli, Cristina Bicchieri, Immanuel Bomze, Peter Hammond, and Brian Skyrms for useful comments and conversations. The usual disclaimer applies.

1. The deep relationships between evolutionary stability and replicator dynamics are illustrated by Bomze and van Damme (1990).
2. Notice that, in the two-strategy case, a monomorphic population state is globally stable under the replicator dynamics if and only if the corresponding strategy is (strongly or weakly) dominant.
3. When punished, defecting players not only have to pay γ but in addition lose the right to receive the reward, which is also equal to γ. This explains why players compare the level of α against that of 2γ rather than simply γ.
4. Since the size of the punishment is given, a norm's strength is measured by the frequency of the punishment.
5. To make the comparison meaningful, I assume that the parameters β and γ are the same for both societies.
6. The assumption of simultaneity of constitutional dates is made for analytical convenience but could be removed. This would require out-of-equilibrium intersocietal comparisons between the "transitional" social wealth of A vs. B, and could further widen the already-rich spectrum of paths of institutional change that is generated by the model and is discussed below in the text.
7. Or "almost" dominate, i.e., move $\hat{\mu}$ (the upper border of the basin of attraction of $\mu = 0$) below μ_0 (which is positive but small).
8. Here and in the following, assume $\alpha < 2\gamma$ unless otherwise indicated.
9. Remember that γ is assumed to be fixed.

REFERENCES

Akerlof, G. A. 1983. "Loyalty Filters." *American Economic Review* **73**:54–63.

Allport, G. W. 1958. *The Nature of Prejudice*. Garden City: Anchor Books.

Arnstein, F., and K. D. Feigenbaum. 1967. "Relationship of Three Motives to Choice in Prisoner's Dilemma." *Psychological Reports* **20**:751–55.

Axelrod, R. 1986. "An Evolutionary Approach to Norms." *American Political Science Review* **80**:1095–1111.

Bicchieri, C. 1990. "Norms of Cooperation." *Ethics* **100**:838–61.

Binmore, K. G., and L. Samuelson. 1992. "Evolutionary Stability in Repeated Games Played by Finite Automata." *Journal of Economic Theory* **57**:278–305.

Blad, M. C. 1986. "A Dynamic Analysis of the Repeated Prisoner's Dilemma Game." *International Journal of Game Theory* **15**:83–99.

Bomze, I. and E. van Damme. 1990. "A Dynamical Characterization of Evolutionarily Stable States." Center Discussion Paper no. 9045, Tilburg University.

Cabrales, A., and J. Sobel. 1992. "On the Limit Points of Discrete Selection Dynamics." *Journal of Economic Theory* **57**:407–19.

Crawford, V. P. 1991. "An 'Evolutionary' Interpretation of Van Huyck, Battalio, and Beil's Experimental Results on Coordination." *Games and Economic Behavior* **3**:25–59.

Dekel, E., and S. Scotchmer. 1992. "On the Evolution of Optimizing Behavior." *Journal of Economic Theory* **57**:392–406.

Elster, J. 1989. "Social Norms and Economic Theory." *Journal of Economic Perspectives* **4**:99–117.

Friedman, D. 1991. "Evolutionary Games in Economics." *Econometrica* **59**:637–66.

Fudenberg, D., and E. Maskin. 1990. "Evolution and Cooperation in Noisy Repeated Games." *American Economic Review* **80**:274–79.

Gilboa, I., and A. Matsui. 1991. "Social Stability and Equilibrium." *Econometrica* **59**:859–67.

Granovetter, M. 1985. "Economic Action and Social Structure: The Problem of Embeddedness." *American Journal of Sociology* **91**:481–510.

Gray, M. R., and S. J. Turnovsky. 1979. "Expectational Consistency, Informational Lags, and the Formulation of Expectations in Continuous Time Models." *Econometrica* **47**:1457–74.

Hofbauer, J., and K. Sigmund. 1988. *The Theory of Evolution and Dynamical Systems*. Cambridge: Cambridge University Press.

Kandori, M. 1992. "Social Norms and Community Enforcement." *Review of Economic Studies* **59**:63–80.

Kliemt, H. 1990. "The Costs of Organizing Social Cooperation." In M. Hechter, K.-D. Opp, and R. Wippler, eds., *Social Institutions. Their Emergence, Maintenance and Effects*. Berlin: de Gruyter, 61–79.

Maynard Smith, J. 1982. *Evolution and the Theory of Games*. Cambridge: Cambridge University Press.

Matsui, A. 1992. "Best Response Dynamics and Socially Stable Strategies." *Journal of Economic Theory* **57**:343–62.

Nachbar, J. H. 1990. " 'Evolutionary' Selection Dynamics in Games: Convergence and Limit Properties." *International Journal of Game Theory* **19**:59–89.

Okuno-Fujiwara, M., and A. Postlewaite. 1991. "Social Norms and Random Matching Games." Mimeo, University of Pennsylvania.

Sacco, P. L. 1994. "Selection Mechanisms in Economics." *Behavioral Science* **39**:311–25.

Samuelson, L. 1991. "Limit Evolutionarily Stable Strategies in Two-Player, Normal Form Games." *Games and Economic Behavior* **3**:110–28.

Samuelson, L., and J. Zhang. 1992. "Evolutionary Stability in Asymmetric Games." *Journal of Economic Theory* **57**:363–91.

Sugden, R. 1989. "Spontaneous Order." *Journal of Economic Perspectives* **4**:85–97.

Swinkels, J. M. 1992. "Evolutionary Stability with Equilibrium Entrants." *Journal of Economic Theory* **57**:306–32.

van Damme, E. 1991. *Stability and Perfection of Nash Equilibria*. Berlin: Springer Verlag.

4

Learning and efficiency in common interest signaling games

DAVID CANNING

Abstract

I examine a common interest signaling game in which players use the empirical frequency of outcomes in a random sample of past behavior to predict the information content of current signals. I show that the long-run behavior of the system converges (with probability 1) to a sequential equilibrium of the game. However, which sequential equilibria are achieved depends on beliefs about signals that have not been sent in the past. If these beliefs are given exogenously, and fixed over time, then any sequential equilibrium can be supported, but if they vary over time, being drawn randomly for every player each period, the system converges to an efficient (fully revealing) equilibrium.

Random beliefs about behavior at unobserved nodes encourage players to make "experiments" and try new messages when signaling is imperfect. These new messages seldom work, but, when by chance they do, imitation by others provides a powerful positive social externality. Exogenous "mistakes," trembles to actions, can play much the same role as "experiments," based on random beliefs, but need not lead to efficiency. The problem is that "mistakes" continue even in a fully revealing equilibrium and can make the fully revealing equilibria fragile. Mistakes lead players to use messages already being used to signal another state; there may be a powerful, and long-lasting, negative social externality when a player causes confusion in this way.

1. INTRODUCTION

Since their introduction by Crawford and Sobel (1982), signaling games have found many applications in economics. One major drawback to their use, however, is that signaling models frequently give rise to a host of sequential equilibria, giving the theory little predictive power. The introduction of refinements, such as requiring the equilibria to satisfy strategic stability in the sense of Kohlberg and Mertens (1986), can reduce the set but often does not lead to uniqueness (see Cho and Kreps [1987] and Banks and Sobel [1987]).

One interpretation of the problem is that signaling games frequently give rise to a problem of language and meaning. If signals are cheap, we can regard them as words and, in order to determine how the players will use these words, we require some assumption as to their conventional meaning. If we assume the game is one of common interest, and the two players speak the same language, it seems plausible to assume that they will coordinate on an efficient outcome; Farrell (1988) and Rabin (1990) work in such a context by assuming that signals have commonly agreed meanings. If players do not speak the same language, it seems unlikely they can convey any information by signaling. Note that in examples for which refinements do lead to uniqueness of equilibria, such as the Spence education model, signals are costly and so "prove" something; with costless signals, refinements based on strategic stability have no bite.

Rubinstein (1991) argues that this multiplicity of equilibria implies that game theory is incomplete and we require the addition of institutional, historical, or environmental characteristics, such as the richness of the language the players speak, which act as selection devices to choose from among equilibria. One approach to this problem is to endogenize the evolution of conventions and institutions, such as language, bringing them into the model. In this view we can regard a society as a dynamic entity, made up of a sequence of games played over time. If players observe the past, they may use the history of recent past behavior, the conventions of the time, to predict the current behavior of other players. Under this interpretation, Nash equilibrium is simply a conventional behavior that is self-enforcing.

This backward-looking adaptive behavior is of course at odds with the traditional approach to game theory, which would prescribe looking for the Nash equilibria of the full dynamic sequence of games. Adaptive behavior relaxes the assumptions of rationality, common knowledge, and infinite computational ability, usually used to justify Nash equilibrium, but it does imply a particular form of bounded rationality on the part of players. The key assumption we require is that players believe their environment to be sufficiently stationary so that recent experience is a good predictor of current behavior. Which modeling strategy is adopted should depend on the facts of the situation being modeled; at present they are best viewed simply as vehicles for speculation.

The evolution of Nash equilibria as conventions in adaptive-behavior models has been studied by Young (1992, 1993); Kandor, Mailath, and Rob (1993), and Kandori and Rob (1992). These papers study learning models in simultaneous-move games in which the steady states are pure-strategy Nash equilibria. The addition of "mistakes" or perturbations to these models results in a subset, often only one, of the steady-state equilibria being selected. Typically all the pure-strategy Nash equilibria are steady states of the system, but the addition of mistakes allows the investigations of the relative stability of these steady states under perturbations.

The aim of this paper is to examine the evolution of language in a signaling model. To do this, I use Young's (1993) model, extended to cover signaling games. New players appear each period, and they use the empirical frequencies observed in a random sample of some finite history of previous plays to predict their opponent's choice. In extensive-form games, we have the problem of what players are to believe at nodes not reached before; specifically, what is the information content of a message never seen before? What is assumed about such beliefs is vitally important; if players assume that the outcome to some choices will be very bad, they may never try these choices and never learn what would actually happen if they did. I make two alternative assumptions about beliefs about signals that have not been tried before: (i) beliefs in such a situation are given exogenously and fixed over time; (ii) beliefs are drawn randomly each period. Assumption (i) seems appropriate if we imagine a fixed population playing the game repeatedly, while (ii) seems more appropriate in an overlapping-generations model, in which the newborn's beliefs need not correspond to those of the old.

We restrict our attention to signaling games with common interest, in the sense that both players get the same payoff for any given pair of states and actions. We show that with fixed beliefs about unobserved signals, the adaptive-behavior system converges, with probability 1, to a steady state that is a sequential equilibrium of the signaling game. In fact, the steady state can be any pure-strategy sequential equilibria, depending on what beliefs agents have about unobserved signals.

With random beliefs about unobserved signals, the system converges, with probability 1, to a fully revealing equilibrium. The random beliefs encourage some players to "experiment." If these experiments are successful, other players copy them and welfare goes up (the game is one of common interest). Eventually the system reaches an efficient, fully revealing equilibrium, then stays there, there being no further incentive to experiment.

An alternative source of experimentation, instead of perturbations to beliefs about unobserved signals, would be to allow the possibility of mistakes when agents choose their signal or action. However, while such a "trembling-hand" model generates a flow toward efficiency, it does not guarantee convergence to a fully revealing equilibrium. The problem is that mistakes continue even after efficiency has been achieved, and the efficient outcomes are not robust to these mistakes.

The papers closest in spirit to what is shown here are those of Matsui (1991) and Warneryd (1990). Matsui shows that in a common interest game, with prior cheap talk, only the Pareto optimal outcomes belong to the unique, cyclically stable set; that is, if population proportions move in the direction of the current best reply, they are eventually limited to efficient outcomes.

The major difference between this paper and that of Matsui is that, in Matsui's paper, actions "off the equilibrium path" have no payoff relevance and, therefore, can be changed arbitrarily. Here actions must be justified by beliefs, even off the equilibrium path.

Warneryd (1990) shows that only the efficient outcomes of a common interest signaling game are evolutionarily stable (i.e., stable with regard to local perturbations). The differences between this paper and Warneryd are twofold. Firstly, I deal with a larger set of games, in which players' payoffs may be nonzero off the diagonal, for which his convergence results do not hold. Secondly, I study a learning dynamic rather than an evolutionary model and emphasize global stability of the dynamic rather than using a local stability criterion.

The results of this paper also emphasize the importance of the source of "mistakes" or random perturbations for convergence results in learning models; different types of perturbations may give different results. The mistakes, or trembling-hand, approach is currently the most common, but it is not obvious that it is the most appropriate. Many different sources of perturbation are possible. In particular a strong case can be made for perturbations in beliefs as the source of variations in behavior.

2. COMMON INTEREST SIGNALING GAMES

Consider a signaling game in which player one is a type s randomly drawn with probability $p(s) > 0$ from some finite type space S (with, say, $\#S$ types) and sends a message m from the finite set M (with, say, $\#M$ messages), where we assume that the message space is larger than the type space ($\#M > \#S$). Player two observes the message m but not the type and must take an action a, from the finite set A, of size $\#A$. In general the payoffs to the two players are given by payoff functions $U_i(s, m, a)$ for $i = 1, 2$.

I restrict my attention to games with common interest in terms of payoffs to types and actions.

Assumption 1

$$U_1(s, m, a) = V(s, a)$$
$$U_2(s, m, a) = V(s, a)$$

Note that we assume that the messages are costless and the payoffs to the two players to a particular state and action are identical. This is more general than Warneryd (1990), who assumes that the number of states and actions are the same and that payoffs are zero to state-action pairs off the leading diagonal.

Now suppose that we have a sequence of players playing the game, $2N$ in each period. In each period, the $2N$ players are matched into N pairs. We suppose that player one in each pair is assigned a type, the allocation of players to types being random but such that a fixed proportion $p(s)$ of players is of each type s. Of course this allocation can be carried out only if $NP(s)$ is integer for each s, which we shall assume. The advantage of this approach, rather than having each player draw his state independently, is that it ensures every state occurs and can be observed in the past.

Before they play the game, each player draws a stratified random sample of pairings of size k from the last T periods, observing the type, message, and action in each. This sample consists of $p(s)k$ observations of a player of each type s, again assuming that k is chosen so that $kp(s)$ is integer (so it is at least 1). These $p(s)k$ observations are drawn randomly (equal probability on each pairing), without replacement, from the $TNp(s)$ pairings in which player one was type s in the last T periods.

Players choose a strategy in the current game so as to maximize their expected payoff, given their beliefs. Beliefs are formed by looking at the empirical frequencies of states and actions, given messages in the past. That is, player one's subjective probability of the action a, given that he sends the message m, is simply the frequency with which player two in his sample has played the action a, when given the message m. Similarly, the subjective probability player two places on player one's being of type s, given the message m, is the number of times players of type s have sent the message m divided by the total number of times m has been sent in the sample.

We can think of players as following a simple algorithm; player one, who knows the state, forms a subjective belief $b(a|m)$, giving the subjective probability of player two's playing the action a, if one sends the message m. Similarly, player two forms a subjective belief $b(s|m)$, giving the probability that player one is type s, given that he sent the message m.

Assumption 2 *Wherever possible, beliefs are formed by*

$$b(a|m) = \frac{\#(observations\ with\ message\ m\ and\ action\ a)}{\#(observations\ with\ message\ m)}$$

$$b(s|m) = \frac{\#(observations\ with\ type\ s\ and\ message\ m)}{\#(observations\ with\ message\ m)}$$

Assumption 3 *If it is not possible to form beliefs by assumption 2 (i.e., if the bottom line is zero), then*

$$b(a|m) = \beta\ (a|m) \quad and \quad b(s|m) = \beta(s|m),$$

where the probability distributions $\beta(.|m)$ are fixed exogenously and are the same for each player.

These assumptions imply a discontinuity in belief formation; if the player has any observations, he uses assumption 2, but without observation he falls back on assumption 3. Note that assumptions 2 and 3 are compatible with Bayesian updating if we assume that players begin with a "flat" Dirichlet prior, which puts equal weight on each state or action. If we assumed such priors in assumption 3, the model would be completely Bayesian. However, I wish to allow a wider class of priors, and therefore impose a discontinuity in beliefs.

Given these subjective beliefs, player one, given his type s, chooses m so as to maximize

$$\sum_{a \in A} V(s, a) b(a|m).$$

Similarly player two, given the message m, chooses a so as to maximize

$$\sum_{s \in S} V(s, a) b(s|m).$$

If several choices give an equal expected payoff, we assume that the player chooses between them according to a strict ordering on each of the sets M and A.

Assumption 4 *Players act so as to maximize their expected utility given their subjective beliefs. If two or more messages give the same expected payoff, senders choose so as to maximize a strict ordering over messages, given by $c(m_i) = i$. If two or more actions for receivers give the same expected payoff, then the receiver chooses so to maximize the strict ordering $c(a_i) = i$.*

The reason for the second part of assumption 4 is that we require a rule for players' behavior in the event of ties. This strict preference is easier to deal with in the model than assuming that players randomize.

Assumptions 1 through 4 completely define the evolution of our random dynamical system. Given the history at time t, our players take their random sample, from their beliefs, and then decide on their messages and actions in period $t + 1$. Dropping the oldest period and adding this new period gives a new history for period $t + 1$. Notice that the state of the system at any time t consists of the messages and actions set in the periods $t - T$ to t. Given this history, we can calculate a probability distribution over the state in period $t + 1$.

In order to begin the process, we assume that we start with an arbitrary history of length T, which players consult at time $t = 0$ and which is updated in the usual way. We denote this initial state as x_0.

Now consider an absorbing state of the system in which each player one's message, given his type, is constant over time and every player two's action, for each message actually sent, is always the same. Note that, in an absorbing state, there is no randomization; each state is associated with a single message and each message with a single action (though several states can use the same message and several messages can produce the same action). An absorbing state is a history of length T at time t that repeats at time $t + 1$ with probability 1.

Theorem 1 *(i) Every absorbing state of the system can be supported as a sequential equilibrium, and (ii) every sequential equilibrium, in which the players use only pure-strategy choices that are strict best replies, given their beliefs, is an absorbing state for appropriate exogenous beliefs.*

PROOF: In appendix.

Since all choices are strict best replies, there is no problem with choosing the "right" degree of randomization. To support sequential equilibria in which players pick from among several best replies would require a precise tie-breaking rule that would depend on the equilibrium selected.

In (i) I show that there exists a sequential equilibrium that supports the messages and actions used in the absorbing state; it need not support the exogenous beliefs the players hold. For example, the exogenous beliefs of player one might be that, for some messages, player two will use a strictly dominated strategy; this belief cannot be supported in a sequential equilibrium.

Theorem 2 *If $T > 2\#S$, $k \leq N$, then for any initial state x_0, the system eventually enters an absorbing state, with probability 1.*

PROOF: In appendix.

The proof uses the same technique as Young (1993). From any initial state there is a possible path of samples that leads in finite time to an absorbing state. The probability of this sample path is small, but finite. At any particular time, it is unlikely that such a path to the steady state occurs, but in the long run, it must occur with probability 1, and the absorbing state is established. My result requires both that the length of memory be long (T is large) and that the population size be large, relative to the sample size players use (N is at least as large as k). These are required for the proof using the sampling scheme I have devised, but it is likely that both these conditions could be relaxed by using a different sampling scheme to reach the steady state.

3. RANDOM "OUT OF EQUILIBRIUM" BELIEFS

Fixed exogenous beliefs can be justified by assuming that the players live forever. These beliefs about messages that have not been sent in the remembered past are unchanging, because the players do not change. However, in a model with finite lives and population turnover, we might imagine that newborns do not have the same exogenous beliefs as old players, for example, in an overlapping-generations model with players living for $T + 1$ periods. Players live and observe play in the game for T periods before it is their turn to play in period $T + 1$. They use the empirical frequencies they observe when young to predict their opponents' play. For messages not sent, we assume their beliefs are arbitrary, but may be different both across and within generations. We replace assumption 3 with 3a, where

Assumption 3a *If beliefs cannot be formed by assumption 2, then the beliefs of sender i at time t are given by $\beta_{it}(a : m)$, which is drawn randomly from an atomless probability measure π_{ma} that has full support on the #A-dimensional unit simplex. Similarly, the beliefs $\beta_{it}(s : m)$ of receiver i at time t are formed by a random draw from an atomless probability measure π_{ms} that has full support on the #S-dimensional unit simplex.*

We assume that all beliefs are possible and that each young player is "endowed" with a belief that is drawn randomly from the set of possible beliefs.

Assumption 5 *Only fully revealing equilibria of the signaling game are efficient.*

Assumption 5 ensures that it matters that signaling reveal the "type" of the sender, the state he is in. Allowing different states to have the same best action, so that it is not necessary to fully reveal the state, complicates the analysis but does not change the efficiency result.

Theorem 3 *Under assumptions 1, 2, 3a, 4, and 5, (i) every absorbing state of the random dynamical system can be supported as a fully revealing sequential equilibrium of the signaling game, and (ii) every fully revealing sequential equilibrium of the signaling game can be supported as an absorbing state.*

PROOF: In appendix.

Theorem 4 *Under assumptions 1, 2, 3a, 4, and 5, provided $T > 1$ and $k \leq N$, the random dynamical system converges in finite time to an absorbing state, with probability 1.*

PROOF: In appendix.

4. THE ROLE OF MISTAKES

We can think of the players trying new messages in section 3 as making "mistakes." They try new messages based on their subjective beliefs about the response these messages will generate, but they will usually be disappointed. It is only the lucky ones, who are matched with a receiver who responds favorably, who receive a higher payoff. More often the payoff to experimentation will be negative. The key point is that the random sampling of the past gives experimentation a positive social externality; if it is successful, it is imitated. The experimenters may, on average, do badly, but their experimentation ensures that the society moves toward efficient behavior.

An alternative approach to experimentation is simply to impose "mistakes" directly by having trembles in the choice of messages and actions. This appears to have a similar "experimentation" quality as mistakes due to random beliefs; however, as we shall see, it has different long-run consequences.

I now return to using assumption 3, instead of 3a, so "out-of-equilibrium" beliefs are fixed once and for all at the beginning of time. As we have already seen, this assumption gives us convergence to a sequential equilibrium of the signaling game but does not lead to efficiency. Suppose, in addition, that in each game the players may make mistakes in their choices. With probability $1 - \alpha$, each player plays his desired strategy, but with probability α he makes a mistake. If player one makes a mistake, he sends a random message, independently of his state, according to the probability distribution σ_M on M. Similarly, if player two makes a mistake, he takes an action randomly, independently of the message sent, according to the probability distribution σ_A on A.

Assumption 6 *Players take their subjective expected-utility-maximizing choices with probability $1 - \alpha$. With probability $\alpha > 0$, they act according to mistake distributions, σ_M on M, or σ_A on A. These mistake distributions have full support.*

A history of outcomes of length T defines a state of the system. Adding the mistakes clearly makes the system ergodic; any state can be achieved given the right combination of mistakes. Let X be the set of states of the system; that is,

$$X = (S \times M \times A)^{TN}.$$

Each game between a pair of players is characterized by a state s, a message m sent by player one, and an action a taken by player two. In T periods, there are TN such pairs, N in each period. Note that so far we have thought of the strategies m and a as the planned (conditional), pure strategies of the players. In the presence of mistakes, each outcome depends on whether or not a mistake has been made.

Theorem 5 *For each $\alpha > 0$, there exists a unique invariant measure u_α on X, such that with probability 1 the empirical frequencies of outcomes converge to u_α. Furthermore, as $\alpha \to 0$, the weight the measures u_α put on the absorbing states of the unperturbed system (with $\alpha = 0$) converges to 1.*

PROOF: By theorem 2, the only invariant measures on the unperturbed system put weight 1 on the steady states. The rest of the theorem follows from Canning (1992).

Adding low-probability mistakes means that the system spends almost all its time in the steady states of the system without mistakes, but may move between these steady states.

Definition *A state, x, is stochastically stable relative to $(\alpha, \sigma_M, \sigma_A)$ if $\lim_{\alpha \to 0} u_\alpha(x) > 0$.*

Clearly, by theorem 3, only the steady states x^* can be stochastically stable. However, we can go further than this. The number of steady states of the system, $\#(x^*)$ say, is bounded above by $\#(x^*) \leq \#M^{\#S}$, where $\#M$ is the number of messages and $\#S$ is the number of states. Since the steady states are pure-strategy sequential equilibria, every player with the same state sends the same message. There are at most $\#M^{\#S}$ assignments of messages to states.

In order to determine which steady states are stochastically stable, we follow the technique of Freidlin and Wentzell (1984). The following definitions are adapted from Young (1993).

Definition *Let x be a state and x' a successor of x in the sense that it is possible to move from x at period t to x' in period $t + 1$, that is, to move in a one-period transition. For each sender in period $t + 1$, his message is a mistake if, given his state, there is no sample in the past T periods for which this message is a best reply. Similarly, for each receiver the action taken is a mistake if there is no sample, from the previous T periods, that generate beliefs for which this action is a best reply given the message seen. The total number of mistakes between x and the successor x' lies between zero and $2N$ (the extremes of no player or every player making a mistake).*

Definition *The resistance $r(x, x')$ between two states is the least number of mistakes in any sequence of one-period transition from x to x'.*

Since only steady states can be stochastically stable, we limit our attention to them. For each steady state x_i^*, consider the set G_i of i-trees in which each steady state is a vertex and it is possible to move from each steady state along a sequence of edges to x_i^*. We say that $(x, x') \in G_i$ if the i-tree to x_i^* has an edge starting at x and ending at x'.

Definition *The stochastic potential of the steady state x_i^* is the least resistance among all i-trees:*

$$v(x_i^*) = \min_{G \in G_i} \sum_{(x,x') \in G} r(x, x').$$

Young (1993) uses the results of Freidlin and Wentzell to show that the stochastically stable states are those that minimize $v(x)$.

Theorem 6 *There exist dynamic systems satisfying assumptions 1–6 for which inefficient equilibria are stochastically stable for arbitrary k and T, satisfying the conditions of Theorem 1.*

PROOF: In appendix.

The key to the proof of theorem 6 is that players may be very sensitive to mistakes even in a fully revealing equilibrium. If a mistake is made, players who are very optimistic about new messages may shift to try these. Failure of the new message can lead to using messages that are already being employed by other types; players try to maximize their own expected payoff, and they do not take into account the negative information externality they impose on others when they use a message, which initially has a clear meaning, for a new purpose.

5. CONCLUSION

In (generic) signaling games with almost common interest, we can justify the fully revealing equilibria on the grounds that they emerge as the globally stable steady state of a simple learning dynamic on the part of the players. One important behavioral assumption is required; beliefs about messages not sent are drawn randomly. The key to getting to the fully revealing equilibria is "experimentation" by players who are optimistic about messages not currently being used. Their optimism may prove false, but when it is successful, imitation by others who see the happy result increases everyone's expected payoff. It proves to be relatively easy to improve the information content of signals; players latch on to new messages very quickly if they work, even if they were initially sent by mistake.

However, if players' beliefs are fixed over time, the system may get stuck in an inefficient equilibrium; players do not try new messages, because of their pessimistic conjectures about the response to new messages. Adding mistakes, a "trembling hand" forces players to try new messages, but this need not lead to efficiency in the long run. Mistakes continue even once efficiency has been achieved, and these can easily force the system out of a fully revealing signaling equilibrium; players searching for a better signal do not take into account the negative information externality they impose on others when they use the same message as someone trying to signal something else.

APPENDIX

Theorem 1 *(i) Every absorbing state of the system can be supported as a sequential equilibrium, and (ii) every sequential equilibrium, in which the players* *use pure-strategy choices that are strict best replies given their beliefs, is an absorbing state for appropriate exogenous beliefs.*

PROOF: (i) In the absorbing state, every player of each type has the same beliefs, since every random sample of the last T periods will be identical, and every player plays the same strategy, since with identical beliefs assumption 4 ensures strategy choices are identical. Furthermore, each player is playing a best reply to the strategy choice of the other type for messages actually sent.

We set the sequential equilibrium messages for player one equal to those used in the absorbing state. For player two, we set the sequential equilibrium actions and beliefs for messages actually sent equal to those used in the absorbing state. For each message m not sent, we select an arbitrary message m' that is sent and set player two's sequential equilibrium belief, given m, identical to the belief given m' and set the sequential-equilibrium strategy choice for m the same as for the message m'.

It is easy to see that the strategies and beliefs we have constructed constitute a sequential equilibrium. Player two's beliefs satisfy Bayes's rule for messages actually sent, while they are consistent for messages not sent; they could be generated by the belief that all deviations to m come from types who play m' in equilibrium. Player two is maximizing given these beliefs. Player one is maximizing given player two's strategy choice; his choice of message is a best reply to player two's strategy for messages actually sent, while for a message m not actually sent it is clear that deviation gives no expected gain over using m', since player two's response will be the same in both cases.

PROOF: (ii) To prove this, we simply set player two's exogenous beliefs about messages not sent equal to those of the pure-strategy sequential equilibrium and set player one's exogenous beliefs about messages not sent equal to the actions prescribed by the sequential-equilibrium strategy of player two. The sequential equilibrium now constitutes an absorbing state of our learning rule. The strategies chosen in the absorbing state give the sequential-equilibrium beliefs under the empirical-learning rule for messages actually sent. For messages not sent, the exogenous beliefs agree with the sequential-equilibrium beliefs. Given these sequential-equilibrium beliefs, the players must play the strict best replies that make up the sequential-equilibrium strategy choices.

Theorem 2 *If* $T > 2\#S, k \leq N$, *then, for any initial state* x_0, *the system eventually enters an absorbing state, with probability 1.*

PROOF: Since $k \leq N$, it is possible for a player to draw his entire sample from just one period. Consider an arbitrary state (a history of length T), x say, at time $t-1$. Suppose, at time t, all players draw the same sample k of pairings on which they base their beliefs, and all these samples come from period $t - 1$. This has a positive probability. Each player will choose a best reply to the beliefs generated by this sample and each will choose the same strategy by assumption 4. It follows that at time t there is a positive probability of obtaining an outcome with choices being identical for senders of the same type and receivers with the same message.

In period $t + 1$ it is possible that all players in state 1 draw identical samples from period t, while all other players draw the same sample from period $t - 1$ as they did at time t. Each will play a best reply to the beliefs generated by strategies seen in these periods. It follows that in period $t + 1$, each player one type 1 is playing a best reply to player two's choice at time t, while player two, and other types of player one, continue playing the same way as at time t.

We now continue to time $t + 2$ and assume that each player in state 1 plays the same as in period $t + 1$ (drawing the same sample from period t), while other player ones play as at time $t + 1$ (sampling from period $t - 1$). All player twos now draw identical samples from period $t + 1$ and play a best reply.

Note that at time t, all player ones of the same type used the same message, and all player twos with the same message played the same action. At $t + 1$ all players act the same as at t, except that players in state 1 get to change their choice, reacting to the current choices of the others. At $t + 2$, each player one plays the same as $t + 1$, but each player two gets to change his strategy.

We repeat this at $t + 3$, holding all other players fixed but letting players in state 2 play a best reply to behavior at $t + 2$. At $t + 4$, player two makes a best reply to behavior at $t + 3$. This continues for $2\#S$ periods, until player one of each type has made a best reply and player two has responded to each. We then repeat this process, letting player one make a best reply, and so on.

We enter a cyclical process from $t + 2\#S + 1$ onward. At time $t + i + 1$, $i + 1$ odd, player one in state $(i/2)$ $[\text{mod } \#S] + 1$ plays a best reply to a sample from period $t + i$. Player one in state $(i/2 + j)$ $[\text{mod}\#S] + 1$ plays a best reply to the sample from $t + i - 2j$ $(j = 0$ to $\#S - 1)$. At time $t + i + 2$, the samples for each player one are from the same periods as at $t + i + 1$. For $t + i + 1$, $i + 1$ odd, player two samples from period $t + i - 1$ and plays a best reply; for period $t + i + 2$, player two samples from $t + i + 1$ and plays a best reply. Between period t and $2\#S$, players sample from the latest of these periods and from period t.

Player two is playing a best reply to the previous period every even period and playing the same as the last period in odd periods. One type of player one changes to a best reply to the previous period in each odd period. In the even periods, and the odd periods in which it is not a type's turn to change, they play the same as they did in the previous period.

Note that a cycle of this process lasts $2\#S$ periods, so we require a memory length T of at least $2\#S$ to guarantee that players can keep a fixed strategy for this long, while other players are changing, until it becomes their turn to change.

Each time player one of some type changes his strategy, he believes it will raise his expected payoff, or if it remains the same, the change must be a move to a message with a higher index $c(m)$. For messages already being sent, this belief will be correct in the period of change, since he correctly predicts player two's pure strategy response to the new message (which is the same as the previous period). Player two then responds in the next period. By keeping his conditional strategy the same as before, player two cannot be worse off than before; player one's expected payoff is at least as large as before, and so, by assumption 1, must be player two's. When player two takes a best response to player one's change, his expected payoff and player one's must be nondecreasing.

It follows that over the two periods either player one's expected payoff has risen, or it has remained the same, but the message he uses in one state has a higher index $c(m)$.

Now consider what happens if player one of type s shifts to a message not currently being employed. He does this because his subjective belief is that it gives a higher payoff, or has the same payoff but a higher index, than his current message. His belief, however, may be wrong, since his exogenously given beliefs may be incorrect, and in the period he makes the switch, his payoff may decline. However, in the next period, player two plays a best response to one's choices. The new message is now identified with the unique state in which it is sent and solicits a best reply that gives the maximum possible payoff to player one in that state. If player one in state s was not getting the best possible payoff before, then his payoff has risen. If he was getting the highest possible payoff, then the reason for the shift must have been that the new message has a higher index, resulting in the same payoff as before with a higher index. Player ones in other states, who send different messages from the one type s moved away from, get the same response as before and are no worse off. However, player ones who send the message that type s moved from may now get a new response. Under the old response, they and player two are no worse off than before. Player two, choosing a best reply, must do at least as well as this, implying their payoff does not fall.

This implies that the expected payoff of player one, averaged over states, either has risen or is the same, but type s has moved to a message with a higher index $c(m)$. Now consider the value function

$$\sum_{s \in S} p(s)[U(s, a(m(s))) + c(m(s))],$$

where s is a type, $m(s)$ is the message sent by that type, and $a(m)$ is the receiver's action, given a message m. This is the expected payoff of player one plus the expected value of his message index. By the argument above, each time one

80

of the types of player one changes his message, this value strictly increases the period after.

Since there are only finitely many states, actions, and messages, this value function can take on only finitely many values. It follows that after some finite time, our process reaches a steady state at which no type of player one wishes to change his message and the same messages and actions are repeated throughout a cycle. Let this steady cycle repeat M times, where $M2\#S > T$.

In the period after this steady-state cycle, the system is in an absorbing state. Consider this state. Memory of length T ensures that every player draws a sample from the steady state. In response to this sample, and their exogenous beliefs, no type of player one wishes to deviate from the strategy employed in the steady-state cycle. In addition, player two, playing a best reply to the strategies employed by player one, will play the same as in the steady-state cycle. This repeats the same messages and actions as used in the previous $T - 1$ periods, which now continue, unchanged forever.

Entering the steady state in this way requires that the cycle of observations occurs as we have set out and is repeated until the steady state is reached. But we have shown this will occur with a finite number of cycles. Such a set of observations over the finite range is unlikely, but at each time t there is a positive probability that it is about to occur. Since the probability of entering an absorbing state within a fixed bounded time is positive at each time t it must eventually occur, with probability 1.

Theorem 3 *Under assumptions 1, 2, 3a, 4, and 5, (i) every absorbing state of the random dynamical system can be supported as a fully revealing sequential equilibrium of the signaling game, (ii) every fully revealing sequential equilibrium of the signaling game can be supported as an absorbing state.*

PROOF: First consider an arbitrary steady state of the system (if one exists). Since this repeats indefinitely, there is a zero probability of leaving it. Suppose it is not fully revealing. It follows that some type of player one, type s say, has a subjective expected payoff (equal to the objective expected payoff) in the steady state, which is less than the maximal possible in a fully revealing equilibrium. Let $V(s)$ be his expected payoff in the steady state and let $U(s, a(s))$ be the best possible payoff given $s(a(s)$ is the best reply given type s). Clearly $U(s, a(s)) > V(s)$ and there exists $q(s) < 1$ such that

$$pU(s, a(s)) + (1 - p) \min_{a \in A} U(s, a) > V(s), \forall q(s) < p \le 1.$$

It follows that there is an open set of beliefs, with strictly positive probability in the random draw, for which the player will want to deviate from the steady-state message given his type s; such beliefs occur with positive probability, and so an inefficient outcome cannot exist in a steady state. By assumption 5, the absorbing state must correspond to a fully revealing equilibrium.

Now consider a fully revealing sequential equilibrium where type s sends message $m(s)$ and player two responds with $a(m(s))$. Since only fully revealing equilibria are efficient, we have that $U(s, a(m(s))) > U(s, a')$, for some a', or else s would not have to be separated from other types for efficiency. Consider a history of length T in which the messages of each type and the actions these induce are those given by the fully revealing equilibrium. This is an absorbing state. Receivers clearly maximize by sticking to their prescribed action, which has a strictly higher expected payoff than switching to any other action. Senders cannot do better by switching to any other message actually being used in the fully revealing equilibrium, since each of these has a strictly lower expected payoff. They can switch to a message not currently being used, if their belief is that it gives a maximal outcome, and is therefore as good as sticking with their prescribed message. But this implies that, for type s to switch to m', we require $p(a' : m') = 0$, which is a zero probability event given his randomly drawn beliefs on the consequences of sending m'.

Theorem 4 *Under assumptions 1, 2, 3a, 4, and 5, the random dynamical system converges, in finite time, with probability 1, to an absorbing state, provided $T > 1$ and $k \leq N$.*

PROOF: There are a finite number of possible orderings of the #M messages in terms of their expected payoffs. It follows that for each type s at least one such ordering has positive probability under the randomly drawn beliefs. There is a positive probability that every type s draws this ordering for messages not currently being sent. If such a draw is made, and repeated, that type will act "as if" he had a fixed exogenous belief about each message, which generated this ordering of expected payoffs.

Consider the sampling scheme in which each sender at time t of each type draws the same random sample from $t - 1$ (which is possible if $k \leq N$) and has the same ordering over messages not sent. It follows that players of the same type will send the same message. Suppose they repeat this message in period $t + 1$ (again sampling from $t - 1$ and drawing the same ordering over messages not sent), while all the receivers at $t + 1$ sample from period t and so send a best reply to messages actually sent at $t + 1$. Clearly this has positive probability.

Consider a type s that does not get a maximal payoff at $t + 1$; let his expected payoff be $V(s)$. Let $a(s)$ be the optimal action for player two given the type s. Now suppose every player one of type s draws a random belief at $t + 2$ that gives the same unsent message, m' say, the highest expected payoff with

$\beta(a(s) : m') = p$ satisfying $p > p^*$, where $p^* < 1$ is defined by

$$p^* U(s, a(s)) + (1 - p^*) \min_{a \in A} U(s, a) \geq V(s).$$

Such a draw has positive probability, since the set of such ps has positive measure. Since $p > p^*$, it is clear that this message is better than any message being sent, each of which generates a pure strategy response that is not maximal. Hence each type not getting a best response switches to a new message. We repeat this behavior for the sender in $t + 3$, allowing him to draw a sample from $t + 1$. Receivers at $t + 3$ draw their sample from $t + 2$, and, since this consists of fully separating messages which are repeated at $t + 3$, we have a fully efficient outcome at $t + 3$. Note that we have only required memory length of at least 2 for this process.

Sampling one period back at $t + 4$ up to $t + T + 2$, we have a history of length T (from $t + 3$ to $t + T + 2$) in which the same fully revealing behavior is repeated at each node. Note that the beliefs about messages not being sent do not disturb this outcome, since, with probability 1, players believe they generate a mixed response that is dominated by the maximal outcome obtained by playing the fully revealing message. From period $t + T + 2$ onward, we are in an absorbing state, and any sample leads to the same set of observations and produces the same outcome, since by theorem 3 the fully revealing outcome is absorbing. Since the sequence of samples and random beliefs needed to move from an arbitrary state at t to an absorbing state at $t + T + 2$ has a positive probability, it follows that the system must eventually enter an absorbing state with probability 1.

Theorem 6 *There exist dynamic systems, satisfying assumptions 1–6, for which inefficient equilibria are stochastically stable for arbitrary k and T satisfying the conditions of theorem 1.*

PROOF: Consider a game with 3 states, $p(s1) = p(s2) = p(s3) = 1/3$, 4 messages, and 3 actions with a matrix of payoffs as follows:

	$a1$	$a2$	$a3$
$s1$	5	0	1
$s2$	0	3	0
$s3$	0	1	3

for which the exogenous beliefs of senders $\beta(a1|m) = 1$ for all m, where $a1$ is the maximal action, given player one is of type $s1$. Further assume that for all messages m, the beliefs of receivers are $\beta(s2|m) = 1$, where the unique maximal action for type $s2$ is $a2$. Now note that $V(s1, a2) < V(s1, a3)$ and that the best reply to the belief

$$b(s1|m) = b(s3|m) = \frac{1}{2}$$

for player two is $a1$ in response to m.

In any efficient steady state (fully revealing) (without loss of generality, set $m(s1) = m1m(s2) = m2$ and $m(s3) = m3$ and $m4$ not used), suppose one player playing as player two makes a mistake against a type-one player one and plays the nonmaximal reply $a2$ to his message $m1$ in period t. This can set up the following pattern of behavior:

time	t	$t+1$	$t+2$	$t+3$	$t+4$
type					
$s1$	$m1 - a1(a2)$	$m4 - a2$	$m3 - a3$	$m3 - a1$	$m3 - a1$
$s2$	$m2 - a2$	$m2 - a2$	$m2 - a2$	$m2 - a2$	$m2 - a2$
$s3$	$m3 - a3$	$m3 - a3$	$m3 - a3$	$m3 - a1$	$m2 - a2$

The entry shows the message sent by the type in that time period and the response of player two. The $(a2)$ term indicates one mistake, a player two playing $a2$ in response to $m1$ in period t.

In $t + 1$ suppose all player ones of type one observe the mistake, $a2$ being played in response to $m1$, and so switch to the message $m4$ not played in the steady state; by assumption this has $\beta(a1|m4) = 1$. Player two responds to $m4$ by playing $a2$, since he believes it is type two ($s2$) sending this new message.

At time $t + 2$, assume all type ones sample once from period t, observing only the mistake response by player two and the rest of their sample comes from period $t + 1$. They now play $m3$, since this gives the highest available expected payoff (1) as opposed to their old message, expected payoff zero, or the new one they just tried, or mixing with type two, both of which also have an expected payoff of zero. It follows that in period $t + 2$, both type one and three play a message $m3$, which produces the response $a3$, while type 2 plays $m2$, which produces $a2$.

At $t + 3$, assume each player one has the same sample as in $t + 2$ and plays as at time $t + 2$. At $t + 3$, player two samples only from $t + 2$, seeing equal numbers of type one and type three playing $m3$ and so responds to $m3$ at time $t + 3$ with $a1$. He responds to type two's message $m2$ with $a2$.

At time $t = 4$, player one samples from $t + 3$, and type $s3$ switches to $m2$; his experience shows $m3$ leads to $a1$ and zero payoff, while his exogenous beliefs indicate that $m1$ and $m4$ will do the same. Type $s1$ stays with $m3$, since it is now maximal, while type $s2$ stays with $m2$. Player two responds to $m3$ with $a1$ and to $m2$ with $a2$.

With one period sampling in the past, the behavior at $t = 4$ continues for T periods and then is an absorbing state. All we require is two or more periods of memory.

It follows from this argument that $r(x^*, x) = 1$, where x^* is our efficient equilibrium, and x is the inefficient steady state we have created. Now note that we can construct an H tree to x by taking the minimal H-tree to x^*, adding a branch from x^* to x, and deleting the branch that leaves x. Since x is a steady state, the branch leaving x has resistance of at least 1. Hence $v(x) \leq v(x^*)$, and if x^* is stochastically stable, so is x. Since for every efficient steady state x^* we can find a corresponding inefficient steady state x with $v(x) \leq v(x^*)$, it follows that at least one inefficient steady state has minimal v and is stochastically stable.

References

Banks, J., and J. Sobel. 1987. "Equilibrium Selection in Signaling Games." *Econometrica* **55**: 647–62.

Canning, D. 1992. "Average Behavior in Learning Models." *Journal of Economic Theory* **57**: 442–72.

Cho, I.-K., and D. Kreps. 1987, "Signaling Games and Stable Equilibria." *Quarterly Journal of Economics* **102**:179–221.

Crawford, V., and J. Sobel. 1982. "Strategic Information Transmission." *Econometrica* **50**:1431–51.

Farrell, J. 1988. "Meaning and Credibility in Cheap Talk Games." In *Mathematical Models in Economics*, M. Demster, ed. Oxford University Press.

Freidlin, M., and A. Wentzell. 1984. *Random Perturbations of Dynamical Systems*. New York: Springer Verlag.

Kandori, M., J. Mailath, and R. Rob. 1993. "Learning, Mutation and Long Run Equilibria in Games." *Econometrica* **61**:29–56.

Kandori, M., and R. Rob. 1992. "Evolution of Equilibria in the Long Run: A General Theory and Applications." University of Pennsylvania, CARESS working paper # 92-06R.

Kohlberg, E., and J. F. Mertens. 1986. "On the Strategic Stability of Equilibria." *Econometrica* **54**:1003–37.

Matsui, A. 1991. "Cheap-Talk and Cooperation in a Society." *Journal of Economic Theory* **54**:245–58.

Rabin, M. 1990. "Communication between Rational Agents." *Journal of Economic Theory* **51**:144–71.

Rubinstein, A. 1991. "Comments on the Interpretation of Game Theory." *Econometrica* **59**:909–24.

Warneryd, K. 1990. "Cheap Talk, Coordination, and Evolutionary Stability." Mimeo, Stockholm School of Economics.

Young, H. P. 1992. "An Evolutionary Model of Bargaining." Mimeo, School of Public Affairs, University of Maryland.

Young, H. P. 1993. "The Evolution of Conventions." *Econometrica* **61**:57–84.

5

*Learning on a torus**

LUCA ANDERLINI, ANTONELLA IANNI

Abstract

We investigate the behavior of some locally interactive learning systems for a finite population of players. We assume that each player is matched with his *neighboring players* to play a coordination game. Players adopt boundedly rational rules of behavior and update their strategy choices with inertia. Learning and experimentation are entirely motivated by payoff considerations. We analyze the asymptotic properties of the aggregate system, and we prove convergence to absorbing states, which we characterize. We report on the results of some simulations carried out on a Cellular Automaton hardware board that simulates a Torus of size 256×256 at a speed of 60 frames per second.

1. INTRODUCTION

Nash equilibrium is certainly at the center of noncooperative game-theoretic analysis. Although used almost universally, its logical foundations have been questioned closely in the recent literature. Within the so-called *eductive* approach, Nash equilibrium is justified on the basis of pure introspection: deductive reasoning on the part of players produces the knowledge of equilibrium strategies that are unerringly identified and played. Refinements of the Nash equilibrium concept provide an attempt to identify a unique plausible outcome in cases of multiplicity. The troubling feature of this approach is that agents are credited with an overabundant capacity for logical reasoning. Moreover, even economists who are relatively comfortable with the assumption of unbounded

*We are grateful to Ken Binmore, Tilman Börgers, Bob Evans, Daniel Probst, Hamid Sabourian, and seminar participants at the Second C.I.T.G. workshop on Game Theory, University College London and Cambridge for stimulating comments. Any remaining errors are, of course, our own responsibility. The CAM-PC board we used for the simulations has been purchased with the support of a SPES grant from the European Community; their financial support is gratefully acknowledged. CAM-PC is © of Automatrix Inc., Rexford, NY.
Address for correspondence: Luca Anderlini, University of Cambridge, Faculty of Economics & Politics, Sidgwick Avenue, Cambridge CB3 9DD, U.K. (e-mail: la13@econ.cam.ac.uk).

computational ability may be puzzled by the amount of coordination required among players: a common prior expectation about how the game should be played must be agreed upon before play begins.

An alternative approach analyzes the *evolution* of play itself: instead of inferring players' behavior from an equilibrium notion, some plausible rules of behavior on the part of agents are postulated and interest is then focused on the evolution of the sequence of play when interaction is repeated over time. Does the process converge? If so, does it converge to a Nash equilibrium? Can equilibrium selection issues be addressed? The idea underlying this kind of model is that players *learn* how to play the game by reacting adaptively to the environment around them. A closely related view draws on the biological literature and relies on the interpretation of payoffs in terms of relative fitnesses. The approach based on learning and evolution can potentially overcome the difficulties we mentioned above, in that it does not place unreasonable demands on players and it focuses on the dynamics of play itself, rather than on that of belief revision. The obvious drawback is the high degree of subjectivity that ensues from the rejection of the (unbounded) rationality-common prior assumption.

The approach we take in this work falls into the latter category, in that we analyze an evolutionary process of learning. A survey of the literature in this field is far beyond the purposes of this introduction; among the growing number of recent contributions, let us recall the surveys of van Damme (1987) and Friedman (1991) and, among others, the works of Binmore (1987) and (1988); Fudenberg and Kreps (1990); Nachbar (1990); Canning (1992); Samuelson and Zhang (1992); Kandori, Mailath, and Rob (1993); Kandori and Rob (1993); Young (1993); and Binmore and Samuelson (1993). Common features of the models that analyze learning in normal form games can briefly be outlined as follows. A population of players are repeatedly and randomly matched to play a one-shot normal form game. At each round, each player decides how to update his strategy choice for the next round on the basis of the information he gathered about the past history of play. Such information generally consists of a sample of outcomes of past play or, typically, of some statistics of how the last round was played. Players are assumed to adopt some boundedly rational "rules of thumb," according to which they decide what to do next. The dynamics of the population as a whole is entirely determined by the intertemporal link stemming from the rules adopted by the players. Interest is then focused on the long-run properties of such a system, and, if convergence obtains, steady states can be characterized.

The full-rationality paradigm is therefore abandoned in favor of behavioral rules that are *myopic*, in the sense of not being forward-looking, and that show some sort of *inertia*, in that not all agents update their strategy choices at the same time. The myopia assumption clearly refers to the fact that agents act in a way that maximizes their expected payoff in the current period of play. The assumption is essential, because the purpose of the analysis is to provide some rationale for equilibrium, and such a possibility would be ruled out almost by definition if fully dynamically optimal behavior were postulated. The inertia assumption reflects the idea that changing strategy may be costly, or that the knowledge that agents have about payoffs or about the strategic behavior of their opponents is at best uncertain. The myopia assumption introduces an explicit path-dependence over strategy choices that governs the dynamics of the system as a whole, and the inertia assumption superimposes a further stochastic component, beyond the one stemming from the random matching among players.

In most of the literature to date, the next step in the analysis is to investigate what happens if the system is constantly hit by small random shocks that are independent across players and over time and have full support in the strategy space. Such random perturbations are meant to model experimentation or mistakes in strategy choices on the part of players and are the precise analog of the basic notion of *mutations* in biological evolutionary theory. While the issue of whether the transposition of biological concepts to economics is still controversial, what is unequivocal is that such a formalization overcomes the limits of the notion of evolutionary stable strategies, in that it allows us to derive a *stochastically stable distribution*, that is, a distribution of strategies that is restored repeatedly when the evolutionary process is constantly buffeted by small random perturbations. The appealing feature of the limit distribution derived under these assumptions is that it is unique and ergodic, and it can be paramaterized in the probability of tremble itself. Any other feature of the model becomes, therefore, irrelevant in the long run, and equilibrium-selection issues can be addressed by letting the limit for such probability become infinitesimally small. It has been proved (Kandori, Mailath, and Rob [1993] and Young [1993]) that for pure coordination games (or, more generally for weakly acyclic games) the equilibria that will be selected are those that show the *minimum stochastic potential*, that is, loosely speaking, the equilibria that require the maximum number of mutations to be upset. Results in this field show that if the underlying game is 2-by-2, then the equilibrium that is reached asymptotically is the *risk-dominant* (in the sense of Harsanyi and Selten [1988]) *equilibrium* of the underlying game, whereas such an analogy cannot be easily generalized to games whose strategy space includes more than two elements.

The work we present in this paper departs from the above-mentioned literature in at least two respects. First, while we fully embrace, and model, the *myopia* and the *inertia* assumption in the behavior of players, we do not postulate any random flow of *mutations*. What we have in mind is that experimentation and trembles are *conditional on the state* in which the system is: on the part of each single player, it is having a switch that actually triggers change itself, and, for the system as a whole, such *noise* ceases to operate once an absorbing state is reached. This amounts to saying that the aggregate system shows a high degree of path-dependence: initial conditions prove to matter in the determination of which absorbing state is reached in the limit.

The second, and most crucial, difference with most of the models within this literature is that we focus on *locally* interactive systems; that is, we assume players to interact directly *only* with a subset of the population. In other words, along the line of Blume (1993), Ellison (1991), and Berninghaus and Schwalbe (1992), we postulate the existence of some exogenous fixed structure in terms of communication in the population, owing to which agents are not influenced by all other agents equally. Nearest players, who we conveniently call *neighbors*, have more influence than those further away. We identify neighbors in terms of spatial location, even though what really matters is the existence of an exogenous communication structure, which may hinge upon similarities among features of players or some specific form of modeled strategic interdependence.

On the grounds of the complexity of the environment, we see the assumption of *bounded rationality* as applicable to our contest of learning how to play a game: each player, confronted with a complex network of relations among players within the population that he only partially recognizes, makes use of some rules of thumb in updating his strategy choices. Such ad hoc rules of behavior are nevertheless consistent with some sort of *reasonable* behavior, in that they are entirely motivated by payoff considerations. To exemplify, consider a generic 2-by-2 pure coordination game. In such a game, the expected payoff from choosing one action is an increasing function of the probability of the opponent's choosing the same action. Loosely speaking, it pays to *imitate* what the majority of the opponents are expected to do. It seems, therefore, to be legitimate to analyze *general majority rules of behavior* (i.e., arbitrary monotonic cut-off rules) when dealing with coordination games, in that they are in this case consistent with myopic best-reply choices on the part of players.

In particular, we will assume that at each moment in time, every player plays the underlying coordination game with *all* of his neighbors and remembers the average payoff that he gets in that round of play. In the following period, when asked to choose one of the actions available, he will compare the above-mentioned average payoff with a *threshold* value (which is itself a function of the payoff structure of the game), and if it is greater than the threshold, he will consider himself to be *satisfied* with what he chose and will stick to the action he took previously. If this is not the case, he *may* instead choose to do something different. Experimentation is entirely driven by payoff considerations, and mistakes are allowed only when the underlying deterministic rule prescribes a change.

The focus on locally interactive systems, beyond adding some realistic features to this highly simplified contest of learning, has some remarkable analytical consequences. First, coordination games, which are (at least weakly) acyclical in the obvious nonlocal analog of our model, can indeed show recurrent behavior in the limit once local interaction is postulated. Second, the steady states of the local system may differ radically from those of the nonlocal analog, in that both strategies can coexist in equilibrium (such a possibility is obviously ruled out if players are assumed to interact with the whole population).

The remainder of the paper is organized as follows. In Section 2, we formally describe the model. In Section 3, we analyze the dynamics of the stochastic system, and we prove the main convergence result. Section 4 reports on some simulation results carried out on a Cellular Automaton machine. Simulations suggest some further analytical findings and allow us to fully characterize the absorbing states. Section 5 informally proposes some economic applications and contains some concluding remarks.

2. THE MODEL

The model is that of a one-shot game, played in periods $t = 1, 2, \ldots$ by a finite population of players. The "back cloth" upon which the dynamics of the game unfolds is given by a *locally interactive structure*, which is entirely determined by a *neighborhood* assignment. In Section 4, this will be characterized in more detail; for the time being, we will stick to a very general formulation.

Let a finite set of sites, N, be given. Each site i is connected to finitely many other sites, $N^i \subset N$. Each site is the address of one player; players who live at any site in N^i are i's *neighbors*. The neighborhood assignment is symmetric, in the sense that $\forall i, j \in N, i \in N^j \Leftrightarrow j \in N^i$, and is homogeneous, meaning that each player is surrounded by exactly[1] m neighbors; that is, $\forall i, \|N^i\| = m$.[2] Given that the set of sites is finite, this amounts to assuming away the need for any boundary conditions.

Despite the high generality of the above characterization, the essential features of any locally interactive systems of this kind are easy to outline. First, the location of players is fixed: a player does not move from his address through time, nor do his opponents. Second, players are assumed to interact directly *only* with their neighbors; any interaction with the rest of the population takes place only indirectly, that is, via neighbors. It goes without saying that such a contest is appropriate to describe a situation in which a player is *not* equally likely to meet every other player in the population; some players (his friends, his colleagues, or his actual neighbors) are more likely than others to be his opponents in the underlying game. We take the extreme version of this and assume that the interaction takes place *only* within the neighborhood. Moreover, there is a considerable overlap in the groups of neighbors, which amounts to saying that the communication structure within the population is indeed complex.

At the beginning of each period, every player i chooses an action (to which he will stick for that round) from the finite set that constitutes his strategy set, S_i, and plays a one-shot normal form game with each of his neighbors. We call this underlying game G and will specify it in more detail later on. We assume that the strategy set is the same for each player (we therefore drop the subscript "i" from the strategy set), and we concentrate on symmetric games. If player i chooses action $a_i \in S$ and player j chooses $a_j \in S$, then $\pi_{i,j}$ denotes the payoff that i gets from the interaction.

If player i were fully rational (and rationality were common knowledge), he would maximize his expected payoff while playing with each neighbor, by taking into account that the action that his opponent chooses depends on his neighbors, and on his neighbors' neighbors, and so on. We take the view that this would put unreasonable demands on the computational ability of players. What we could assume, instead, is that players act *myopically*, in the sense that they believe the distribution of actions within their neighborhood is the same as they observed in the previous period. The best they can do, then, is to choose the *myopic best reply* to it:

$$a_{i,t}^{MBR} \in \arg\max_{a_i \in S} \ \pi_i(a_{i,t}, a_{-i,t-1}),$$

where a_{-i} refers only to (and to all of) i's neighbors,[3] assuming an arbitrary tie-breaking rule is applied whenever such a best reply is not unique.

Clearly, the assumption of myopia introduces a deterministic intertemporal link over strategy choices, which we could conveniently define as *local best-reply dynamics*. A stochastic component could be superimposed on such dynamics in the following way. Consider a player who has chosen a_i at t, and suppose the myopic best-reply rule of behavior, described above, tells him to switch to a_j at $t + 1$. Then his behavior would show some sort of *inertia* if he does not necessarily follow the prescription: if, for example, he sticks to the action he adopted at time t with some positive probability. This is the line we take in Anderlini and Ianni (1995).

In the present work, we concentrate on a slightly different specification of the behavioral rule (that, for specific classes of games, we prove to be equivalent). Focusing on the myopic best reply requires players to remember some statistics of the actions played within the neighborhood in the previous period. Sometimes such knowledge might be difficult, or costly, to collect and retain. We therefore prefer to think of our player as being less informed than the modeller. In particular, we assume that, at the end of period t, after having played G with all his neighbors, player i remembers only the average payoff he obtained in that round of interaction. The average payoff for player i at t is obviously a function of his action and the actions taken within his neighborhood; we denote it by

$$\overline{\pi}_{i,t} = \overline{\pi}_i(a_{i,t}, a_{-i,t}).$$

We then focus on the following class of stochastic behavioral rules:

Assumption 1 *Given $a_{i,t}$ and $\overline{\pi}_{i,t}$, player i will revise his strategy choice at the beginning of period $t + 1$ according to an adjustment rule for which the following holds:*

$$\Pr(a_{i,t+1} = a_{i,t}) = 1 \quad iff \quad \overline{\pi}_{i,t} > \tilde{\pi}_i$$
$$\Pr(a_{j,t+1}) > 0 \quad \forall j \quad if \quad \overline{\pi}_{i,t} \le \tilde{\pi}_i$$

where $\tilde{\pi}_i$ is a conveniently chosen deterministic value that is a function of the payoffs of G.

This general specification formalizes the idea that players compare what they actually got with a threshold that represents their *aspiration level*.[4] If the payoff exceeds such threshold, then they will consider themselves to be satisfied with the action they took, they will not change it in the next period, and they will therefore be referred to as *stable* for the purposes of our analysis. If this is not the case, a player might instead decide to do something else, adopting any feasible action with positive probability. A player in such a situation will be defined as *unstable*. Note that the definition of stability that is implied by Assumption 1 is entirely defined by what happens *within* the neighborhood.

Despite the highly general formulation, a number of considerations are to be noticed. First, in this framework, it is exclusively experience that drives strategy revision: whenever a player is happy with the action he took, there is no reason whatsoever for him to switch to a different strategy. The immediate implication of such an assumption is that once a player is adopting an equilibrium strategy, he will not abandon it.[5] On the contrary, a player who is not satisfied with the payoff he got in the previous round of interaction leaves the door open to *any* possible action. He may, therefore, stick to what he had done before, or he may as well switch to a different strategy. By adopting a completely mixed strategy, he certainly does not rule out the possibility of choosing the myopic best reply to his neighborhood (even though he may not be aware of it), and his behavior may accordingly show some *inertia*. Stated somewhat differently, we would like to think of very simple rules of behavior as the "primitives" of our analysis and investigate whether they might constitute a rationale for an equilibrium notion.

We now turn to the specification of the underlying game, G. Play occurs more frequently than strategy revision: in any period, t, every player meets each of his m neighbors and plays a one-shot normal form symmetric n-by-n game, G, which is characterized by a payoff matrix of the type:

$$
G = \begin{bmatrix} \pi_{1,1} & \pi_{1,2} & \cdots & \pi_{1,n} \\ \pi_{2,1} & \pi_{2,2} & \cdots & \cdots \\ \cdots & \cdots & \cdots & \cdots \\ \pi_{n,1} & \cdots & \cdots & \pi_{n,n} \end{bmatrix}
$$

We take G to be a *pure coordination game*, in the sense that the following holds:

Assumption 2 G *is such that*

$$
\pi_{i,i} > \pi_{j,i} \quad \forall j \neq i \quad i, j = \{1, 2, \ldots, n\}.
$$

Assumption 2 formalizes the idea that it is always optimal to play the same strategy as one's opponent,[6] and it has the obvious implication that all the outcomes on the main diagonal are Nash equilibria of G.

The notion of *risk dominance* has taken a central role in the literature concerning coordination games. Introduced by Harsanyi and Selten (1988) for 2-by-2 games, it can be stated as follows:

Definition 1 *Define i and j to be the equilibria where both players adopt strategy i and j respectively. Then i weakly risk-dominates j, which we denote by $(a_i, a_i) \overset{r}{\succeq} (a_j, a_j)$, if and only if*

$$
\pi_{i,i} - \pi_{i,j} \geq \pi_{j,j} - \pi_{i,j} \quad i \neq j.
$$

As is known, a risk-dominant equilibrium may or may not be dominant in Pareto terms. Moreover, it should be noted that, if $n > 2$, such a definition allows only for *pairwise comparisons*, in that, while ranking (in terms of risk) equilibrium i with equilibrium j, no attention is paid to *any* other strategy $j' \neq i, j$. Clearly, if G is 2-by-2, such a caveat does not apply. Finally note that, under Assumption 1 on G, an equivalent formulation of Definition 1 is the following:

$$
(a_i, a_i) \overset{r}{\succeq} (a_j, a_j) \quad iff \quad r_{i,j} = \frac{\pi_{i,i} - \pi_{j,i}}{(\pi_{i,i} - \pi_{j,i}) + (\pi_{j,j} - \pi_{i,j})} \geq \frac{1}{2}.
$$

In what follows, we will define $0 < r_{i,j} < 1$ to be the risk coefficient associated with the choice of strategy i vs. strategy j. Note that $r_{i,j} = 1 - r_{j,i}$.

In most of this paper, we deal with 2-by-2 coordination games. This simplifies the analysis in a number of respects. First, the strategy space of each player is a Boolean space. Clearly, this amounts to saying that knowing that a player "switches strategy" entirely defines what he will do next. Second, as we will show below, for any coordination game, the adjustment rule can equivalently be formulated in terms of *majority rule of behavior*.

To start with, consider the following very simple specification of the underlying game:

$$G_1 \equiv \begin{bmatrix} 1 & 0 \\ 0 & 1 \end{bmatrix}$$

Suppose today a player, who played G_1 with his m neighbors yesterday, decides to update his strategy choice following a rule of thumb of this kind: "At time t, do what the majority of your neighbors have done at $t - 1$." Would this behavior be consistent with some sort of myopic optimization? Could this simple behavioral rule be expressed in terms of an aspiration level in terms of payoff considerations? The answer to the latter question is clearly yes: an entirely equivalent specification of the rule would run as follows: "*If, at time $t - 1$, you achieved an average payoff of at least $m/2$, then stick to what you did at t. If not, then do the opposite.*" It also happens to be the case that, by following this simple rule of thumb, our player would behave *as if* he was adopting the myopic best reply to his neighboring players. In other words, for G_1, myopic optimizing behavior and simple majority rule are exactly equivalent. Could this equivalence between simple *monotonic cut-off rules*, *payoff-memory* considerations, and *myopic optimization* be generalized to any 2-by-2 coordination games? An answer is provided in Remark 1.

Remark 1 *Consider a 2-by-2 coordination game for which Assumption 1 holds. Assume $(a_i,\ a_i) \overset{r}{\succ} (a_j,\ a_j)$, or, equivalently, $r_{i,j} > \frac{1}{2}$. Define $m_{i,t}$ to be the function of neighbors who adopted strategy a_i at t ($\sum m_i = 1$). Then the rule of thumb behavior*

$$a_{i,t+1} = a_{i,t} \quad \textit{iff} \quad m_{i,t} > \tilde{m}_i = 1 - r_{i,j} \quad i, j = \{1, 2\} \quad j \neq i$$

is exactly equivalent to

$$a_{i,t+1} = a_{i,t} \quad \textit{iff} \quad \overline{\pi}_{i,t} > \tilde{\pi} = \pi_{i,i}(1 - r_{i,j}) + \pi_{i,j}r_{i,j} \quad i, j = \{1, 2\} \quad j \neq i$$

and is consistent with myopic best-reply behavior on the part of the player. Note that, if $(a_i, a_i) \overset{r}{\sim} (a_j, a_j)$, then $r_{i,j} = \tilde{m}_i = \frac{1}{2}$, and $\tilde{\pi} = (\pi_{i,i} + \pi_{i,j})/2$.

The proof is trivial and is therefore omitted. The equivalence is clearly due to the fact that $1 - r_{i,j}$, beyond quantifying the risk associated with choosing one strategy instead of the other, also represents the fraction of neighbors playing i that would make a player *exactly indifferent* between the two strategies. It goes without saying that, in 2-by-2 games, the risk coefficient corresponds to the mixed-strategy Nash equilibrium of the underlying game.

An immediate, and relevant, implication of such specification of the adjustment rule is that, in a 2-by-2 game, a player who adopts it is unequivocally adopting the *myopic best reply* to his neighborhood and, therefore, the *Nash equilibrium* strategy if the distribution of strategies had to be stationary, that is, in a steady state. This is due to the general consideration that for 2-by-2 coordination games the basin of attraction of an equilibrium is exactly equivalent to the best-reply region for the strategy that, if played by all players, yields that equilibrium. The latter argument breaks down if the underlying game is an *n*-by-*n* coordination game: switching strategy, if the strategy space is *not* a Boolean space, does *not* necessarily mean that the player will adopt the myopic best reply to his neighborhood. In order for him to choose it, we must, in general, endow him with the information he needs to identify it.

There is a particular case in which a straightforward extension to 3-by-3 games is possible. Loosely speaking, if there are no relevant differences in the risk associated with each strategy, then rules of behavior that require a minimum amount of information, or computational ability, on the part of players, turn out to be consistent with best-reply behavior. This is formally stated in the following Remark 2.

Remark 2 *Consider a 3-by-3 coordination game for which Assumption 2 holds, and that, furthermore, has a unique completely mixed Nash equilibrium. Assume that the three Nash equilibria in pure strategies are risk-equivalent, that is, $r_{i,j} = \frac{1}{2}$ $\forall i, j$ $i \neq j$. Let $m_{i,t}$ represent the fraction of neighbors who adopted strategy a_i at t ($\sum m_i = 1$). Then the rule of thumb behavior*

$$a_{i,t+1} = a_{i,t} \quad iff \quad m_{i,t} > \frac{1}{3} \quad \forall i = \{1, 2, 3\}$$

is exactly equivalent to

$$a_{i,t+1} = a_{i,t} \quad iff \quad \overline{\pi}_{i,t} > \tilde{\pi}_i = \frac{1}{3} \sum_j \pi_{i,j} = \tilde{\pi} \quad \forall i, j = \{1, 2, 3\}$$

and yields a behavior that is consistent with myopic best-reply behavior on the part of players.

Clearly, for any coordination game, the formulation in terms of aspiration level of payoff, in Remarks 1 and 2, is the *quiet* version of the general class of noisy rules identified by Assumption 1. The former specifies the deterministic dynamics for the whole population of players, and the latter introduces a stochastic component that stems from the assumption of inertia. The interest is therefore focused on the behavior of such dynamics in the limit. These are the issues addressed in Section 3.

3. CONVERGENCE RESULTS

In this section, we analyze the dynamics of the process that defines the evolution of strategy choices for the *whole* population of players. As stated previously, we deal with a finite population of N players; the state of the system, θ, is identified by an N-tuple $(a_1, a_2, \ldots, a_N) \in \Theta \equiv S^N$, where Θ is the state-space. Given the local nature of the interaction, the vectorial notation for θ allows us to keep track of the "addresses" of each player, as well as of the strategy he chooses: the system cannot unambiguously be tracked by any statistics of the population, such as (as is customary for nonlocal interaction models) the total number of players playing each action.

The dynamics can be thought of as a discrete iteration of a function, B, on the finite set of states, Θ:

$$\theta_{t+1} = B(\theta_t).$$

B is a mapping from the set of states, Θ, into itself, that defines the one-step adaptation of the strategy choices for the whole population of players. An intuitive way of taking into account the identities of players is to consider B as a composite function of each single b_i:

$$B(\cdot) = (b_1(\cdot), \ b_2(\cdot), \ldots, \ b_N(\cdot)),$$

where each $b_i : S^m \to S$ is a mapping that is entirely defined by the adjustment rule that we postulate. In general, we will focus on adjustment rules that are *homogeneous*, in the sense that each player has exactly the same strategy set and the same transition rule applies, and are *local* in both space and time, in that the next action he takes depends only upon the current action of his neighbors and of himself. Moreover, we will distinguish between *deterministic* and *stochastic* adjustment rules, the latter being those for which the action actually chosen depends on the previous configuration within the neighborhood, as well as on the determination of a random variable.

An equivalent interpretation of the transition function $B(\cdot)$ is in terms of *parallel* modes of operation of the associated *automata network* (see Robert [1986] for a formal analysis of these issues). Every player, located at a vertex of the connectivity graph defined by the neighborhood assignment, can be interpreted as a finite-state automaton: at any step t, all automata adapt simultaneously. We will follow this line while dealing with the simulation results. For the time being, we simply point out that the two interpretations are exactly equivalent.

Investigating the limit behavior of the dynamic system involves looking for fixed points of B. Owing to the finiteness of the domain and range of B, given an initial condition, θ_0, the sequence of iterations may converge either to a *fixed point* or to a *limit cycle*. Following the dynamics of B is quite a complex task: one possibility is to visualize the discrete iteration as the iteration graph for

B, that is, a graph whose vertices are the n^N possible states and whose edges connect each θ with $B(\theta)$. Unless numbers are very small, this approach is not really operative. An approximation of B is given by the Gauss-Seidel operator: the Gauss-Seidel iteration for B is defined to be the *serial* mode of operation, in which all players, with fixed numbering, adapt sequentially:

$$\mathcal{G} = B_N \circ B_{N-1} \circ \ldots \circ B_1,$$

where each B_i has N components: $B_i = (I(\cdot), \ldots, b_i(\cdot), \ldots, I(\cdot))$.[7] This constitutes a more tractable object, and, moreover, it has been proved (see Robert [1986]) that B and its associated Gauss-Seidel approximation have the *same* set, possibly empty, of fixed points (nothing can be said about limit cycles).

If the underlying coordination game is 2-by-2, the strategy space is a Boolean space and the function b_i is, accordingly, a Boolean function: $b_i : \{0, 1\}^m \to \{0, 1\}$. Dealing with such objects has some technical advantages: one could introduce a (Boolean) metric and hence be able to use real analysis tools to investigate the limit properties of B. This is, in fact, what is done in Berninghaus and Schwalbe (1992) for an entirely deterministic system. Their findings show that global convergence requires very restrictive conditions, and, in most cases, only local convergence, in a sense to be specified, obtains.

In this paper, we deal with a stochastic system, in that we postulate b_i to be a stochastic adjustment rule that maps S^m into a *subset* of S, according to a probability distribution, implicitly defined in terms of some properties it has to satisfy. Given the finite memory of the adjustment rule, the dynamical system defines a Markov chain on the finite state-space Θ. We may define the time t probability distribution over the states by an n^N-dimensional vector d_t; the process is therefore governed by

$$d_{t+1} = P d_t,$$

where d_t belongs to the n^N-dimensional simplex, and P is the transition matrix, the elements of which represent the probability with which the stochastic process transits from one state to another. Given the neighborhood structure of the interaction, P acts on n^N-dimensional vectors, and it is difficult to visualize its structure.

Investigating the asymptotic behavior of the system is essential to the purposes of our analysis: in order to provide a rationale for equilibrium, we would like, first, for the process that describes the evolution of play to converge to an absorbing state (limit cycles would be of difficult interpretation) and, second, to be able to characterize such steady states in terms of an equilibrium notion. Along this line, we are now ready to state the first convergence result. The underlying game will be assumed to be a 2-by-2 coordination game, and, for notational convenience, the two actions available to each player will be relabeled 0 and 1.

Theorem 1 *Consider a finite population of N players, each of them repeatedly matched with his m-neighbors to play a 2-by-2 coordination game that satisfies Assumption 2. The neighborhood assignment is symmetric and homogeneous. Assume each player revises his strategy choices according to a rule consistent with Assumption 1, restated formally as follows: for all i, b_i is a probability distribution on $\{0, 1\}$ such that*

$$\Pr(a_{i,t+1} = a_{i,t}) = 1 \quad \textit{iff} \quad \overline{\pi}_{i,t} > \tilde{\pi} = \pi_{i,i}(1 - r_{i,j})$$
$$+ \pi_{i,j} r_{i,j} \quad i, j = \{0, 1\} j \neq i$$

$$\Pr(a_{i,t+1} = a_{i,t}) = \varepsilon > 0 \quad \textit{iff} \quad \overline{\pi}_{i,t} \leq \tilde{\pi}.$$
$$\Pr(a_{i,t+1} = 1 - a_{i,t}) = 1 - \varepsilon$$

Then, (a) the stochastic system

$$d_{t+1} = P d_t,$$

given any initial condition θ_0, will converge in finite time to an absorbing state, and (b) steady states are Nash equilibria of the N-player game, in the sense that each player is adopting the best reply to his neighborhood. Formally,

$$a_i \in \arg\max_{a_i \in S} \pi_i(a_i, a_{-i}),$$

where $-i$ refers only to (and to all of) i's neighbors.

PROOF: In order to prove part (a) of Theorem 1, we have to show that the process is weakly *acyclical*, in the sense that all states communicate in finitely many steps with an absorbing state. Recall that unstable players might switch action between t and $t+1$: due to the inertia assumption, both available strategy choices support of the probability distribution that defines the behavior of an unstable player. On the other end, stable players *do not* switch action. In other words, the probability distribution that defines the transition rule does *not* necessarily have full support. This amounts to saying that the stochastic component that stems from the inertia assumption introduces new links between states that would not communicate in the deterministic system, but it is not *enough* to guarantee regularity of the transition matrix. In other words, what happens in the limit does indeed depend on the initial condition.

For a fixed t, consider a state θ, and define $k \geq 0$, the number of *unstable* players. Specify $0 \leq k_1 \leq k$ to be the unstable players who are choosing action 1. Assume $k > 0$ (if not, θ would already be absorbing).

Imagine you could number the players in a fixed way and think of them adjusting their strategy choice *one at a time*. This is obviously an approximation of the actual evolution of the system between t and $t+1$ and amounts to decomposing the dynamics in single virtual steps: at each step, only one player is allowed to revise his choice. In a (less imaginative and more) formal way, by doing this the function B is approximated by the Gauss-Seidel operator, $\mathcal{G}$. Obviously there are $N!$ possible serial modes of operation for B, and therefore $N!$ $\mathcal{G}$ operators. Furthermore, notice that, given the formulation of the adjustment rule, an unstable player who, ceteris paribus, switches action, becomes necessarily a stable player. This is due to the fact that he will conform to a *majority* (in a broad sense) of neighbors, thereby coordinating his action with them.

Follow the evolution of the system along the following one-step-at-a-time path. Choose a strategy, say 1, and imagine that all k_1 unstable players playing that strategy do actually switch action. After the flip, each of them will necessarily be a stable 0-player. The total number of stable 0-players will therefore be strictly increasing. k_1 may increase along the path, but, in a finite number of steps, the system will necessarily be in a new state, θ', such that either all players in the population are choosing action 0 (i.e., an absorbing state in which strategy 1 has disappeared), or all 1-players are stable. Assume the latter is true; that is, suppose $k > k_1 = 0$. Then repeat the same procedure with the unstable 0-players. By flipping each of them one at a time, the total number of 1-players in the population will be strictly increasing. You *cannot*, therefore, create new unstable 1-players. Owing to the finiteness of the population, we must necessarily reach a state in which either all the population consists of players choosing the same action, or all players in the population are stable and none of the strategies has disappeared. In both cases, the system has reached an absorbing state.[8]

In order to prove part (b) of the theorem, recall that, as pointed out in Remark 1 in Section 2, in a steady state, player i is surrounded by at least $m(1 - r_{i,j})$ neighbors choosing action i. Given the definition of $r_{i,j}$, this trivially amounts to saying that he is maximizing his expected payoff from that round of interaction, which in turn implies that he is choosing the *best reply* to his neighborhood. Owing to the local structure of the interaction, this is enough to show that claim (b) holds. Q.E.D.

Theorem 1 shows that, under an adjustment rule of the kind specified in Assumption 1, the stochastic system that governs the evolution of strategy choices in the population, given an initial condition, converges to an absorbing state in finite time with probability 1. In other words, assuming that unstable players may switch with positive probability ensures that the process cannot get stuck in a limit cycle. Clearly the stochastic component introduced in the system with the inertia assumption does not select among different equilibria, nor does it allow us to define a probability distribution over the set of steady states: which of the absorbing states will be reached depends entirely on the initial condition.

Moreover, the neighborhood structure of the interaction allows for "mixed" configurations in the limits: in equilibrium, different strategies can survive within homogeneous areas. If the matching were nonlocal, such configurations would only be transient, in that, in a steady state, each player would necessarily be adopting the same (best-reply) action. A wide variety of equilibrium configurations are, therefore, possible; we will characterize them more in detail in Section 4, with the help of simulation results. The intuition behind the latter analytical findings is that, in a locally interactive contest, players can be sufficiently inward-looking to coordinate their strategy choices with their neighbors, regardless of what happens outside their neighborhood.

Note that the main argument that drives the proof is that, under the stated assumption, a player who switches action becomes necessarily stable. Loosely speaking, given the underlying payoffs, the notion of stability (defined on the basis of simple majority rules of behavior) turns out to be equivalent to adopting best-reply behavior (that we do not postulate by assumption). This allows us to follow the process along a path, which will be taken with positive probability, and leads the system to an absorbing state that proves to be a Nash equilibrium. In 3-by-3 games this is, in general, no longer the case, in that adopting a best-reply behavior does indeed require more sophisticated computations. For generic n-by-n games, such argument becomes even more plausible. We find that the straightforward extension of the result to 3-by-3 games is possible only in the specific case in which, loosely speaking, risk considerations do not matter. This is formally stated in Theorem 2.

Theorem 2 *Consider a finite population of N players, each of them repeatedly matched with his m-neighbors to play a 3-by-3 coordination game that satisfies Assumption 1 and that has a unique, completely mixed Nash equilibrium. Furthermore, assume that G is such that $r_{i,j} = 1/2$ $\forall i, j$ $j \neq i$, that is, that the 3 equilibria are risk-equivalent. The neighborhood assignment is symmetric and homogeneous. Assume each player revises his strategy choices according to a rule consistent with Assumption 1, restated formally as follows: for all i, b_i is a probability distribution on $\{0, 1\}$ such that*

$$\Pr(a_{i,t+1} = a_{i,t}) = 1 \quad \text{iff} \quad \overline{\pi}_{i,t} > \tilde{\pi} = \frac{1}{3} \sum_{j=1,\dots,m} \pi_{i,j} \quad j \neq i.$$

$$\Pr(a_{i,t+1} = a_{i,t}) = \varepsilon > 0 \quad \text{iff} \quad \overline{\pi}_{i,t} \leq \tilde{\pi}.$$
$$\Pr(a_{i,t+1} \neq a_{i,t}) = 1 - \varepsilon.$$

Then, (a) the stochastic system

$$d_{t+1} = P d_t,$$

given any initial condition θ_0 will converge in finite time to an absorbing state, and (b) steady states are Nash equilibria of the N-player game, in the sense that each player is adopting the best reply to his neighborhood; that is,

$$a_i \in \arg\max_{a_i \in S} \pi_i(a_i, a_{-i}),$$

where $-i$ refers only to (and to all of) i's neighbors.

PROOF: The proof is a straightforward extension of the proof of Theorem 1. Consider an initial θ_0 nonabsorbing. Choose a strategy, say a_1. Flip, one at a time, each player choosing it, to the action that has been adopted by the *majority* of his neighbors (i.e., flip 1 to $j \neq 1$ such that $m_{j,t} > \frac{1}{3}$.) By doing this, he will necessarily become a stable j-player. It must be that in a finite number of steps, you reach a state θ_1 in which there are no unstable 1-players left. Assume θ_1 is not absorbing, then repeat the procedure with a_2, flipping each player to a stable $k(\neq 2)$-player. Note that the latter player *cannot* become an unstable a_1-player. Repeat the procedure for any unstable a_3-players. Given the finite population of players, this procedure must eventually lead to an absorbing state.

The proof of part (b) relies on exactly the same argument as the proof of Theorem 1 and therefore is omitted.

4. LEARNING ON A TORUS

In the previous sections, we dealt with the issue of local learning from a highly abstract point of view. In this section, we adopt a more concrete approach. The result stated in Theorem 1 ensures that, in the limit, the system will converge to an absorbing state and points out that local learning rules may yield a wide variety of steady states that are radically different from those of the nonlocal analog. To characterize the steady states of the system, we rely on the results of simulations that we carried out using a "Cellular Atomaton machine."

The prototype of what is termed a Cellular Automaton[9] is informally described by the following set-up. Consider a finite chessboard in which each square can be one of n colors at each moment in time. Assume that we have a rule that specifies what the color of each square should be as a function of its four *neighboring* squares – north, south, east, and west. Now let an initial pattern of colored squares be given at time $t = 0$. We then turn the system on and let the rule-of-state transition operate, examining the pattern that emerges as $t \rightarrow \infty$. In other words, we watch the squares taking different colors at each time step in accordance, and look for the steady-state pattern, if any, that emerges as t becomes large. It is intuitively clear that the key ingredients of this recipe are exactly those we were using in Section 2. In particular, our requirements in terms of *homogeneity* and *locality* are met: each cell of the system is the same as any other (in the sense that cells can each take on exactly the same set of n colors), and the transition rule is local both in space and time (in that we do not have any time-lag effects, nor do we have nonlocal interaction affecting the state). More accurately, this setting describes a two-dimensional cellular automaton, since the state space is a planar grid. There are, traditionally, two basic neighborhoods of interest: the von Neumann neighborhoods (those of the example above), and the Moore neighborhoods (which also include those cells that are diagonally adjacent).

We are now ready to describe in more detail the set-up of our simulations. We assume our N players to live on a 2-dimensional lattice of 256×256 cells, each cell constituting one player's address. Each player interacts repeatedly *only* with his eight closest neighbors; north, northeast, east, southeast, south, southwest, west, and northwest. The order with which the interaction takes place is irrelevant for the purposes of our analysis, in that what matters is what each player gets, on average, from the interaction. What is formally relevant is that we deal with Moore neighborhoods (which, loosely speaking, guarantees symmetry in any direction) and that the cardinality of the neighborhood is fixed and equal to 8. Moreover, we assume our players inhabit a world in which there are no boundaries: instead of making explicit provisions for genuinely dealing with boundaries (where players would inevitably be surrounded by fewer than 8 neighbors), we assume some sort of periodic boundaries. More precisely, we assume that the square grid is folded (first, the upper edge glued with the lower, and then the right one with the one on the left) to form a geometric object, known as a *Torus*.

When a player interacts with one of his neighbors, he plays the one-shot normal form coordination game, G. At the end of each time period, he averages the payoffs he got in the 8 interactions, and, in the next period, the latter value is all that he remembers of that round of interaction. Having to choose what to do next, he will compare this with his *aspiration level*, $\tilde{\pi}$. Recall that such threshold value is a *deterministic* linear function of the payoffs of G, and our player, remarkably stable in his mood, will keep this as a behavioral reference in *every* period.[10] Remark 1, in Section 2, showed that any 2-by-2 coordination game that satisfies Assumption 1, is, in terms of the dynamics we postulate, in some respects equivalent to the following "simplest" version:

$$G_1 \equiv \begin{bmatrix} 1 & 0 \\ 0 & 1 \end{bmatrix}.$$

More precisely, for any underlying 2-by-2 coordination game, the payoff-memory rule we assume yields the same dynamics as a *totalistic*[11] rule applied to G_1: what a player will do depends on payoff considerations ($\overline{\pi} \geq \tilde{\pi}$), and these are, in turn, determined by the number of neighbors who have chosen the same action as he has $(m \geq \tilde{m})$. Therefore, what the player is implicitly remembering is the number of neighbors with whom he could coordinate his strategy choice. This is clearly equivalent to assuming the underlying game being G_1 and the following local (totalistic) rule of behavior. Identify the address of player i by the coordinates k and l. Assume Moore neighborhoods so that

$$\forall i \quad -i = j \in \{(k, l+1), (k, l-1), (k+1, l), (k-1, l),$$
$$(k+1, l+1), (k-1, l-1), (k+1, l-1), (k-1, l+1)\}.$$

Then, an equivalent formulation of the class of rules stated in Assumption 1 is the following:

$$\Pr(a_{i,t+1} = a_{i,t}) = 1 \qquad iff \quad \sum_{i=1,\dots,8} |a_{i,t} - a_{j,t}| < \tilde{m};$$
$$\Pr(a_{i,t+1} = 1 - a_{i,t}) = 1 - \varepsilon > 0 \quad iff \quad \sum_{i=1,\dots,8} |a_{i,t} - a_{j,t}| \geq \tilde{m},$$

where actions have been conveniently relabeled 0, 1. By rescaling the cut-off value, we can therefore consider any underlying game.

We started by considering $\tilde{m} = 4$. The dynamics of the simulations are not difficult to describe in words. Consider a small cluster of players playing one strategy; the players on the "edges" are necessarily *unstable*, that is, not surrounded by a majority of players choosing the same action they do and, therefore, likely to switch. The small cluster will thus be progressively "eroded" and eventually disappear. Which are the mixed configurations of strategy choices that are stable, in the trivial sense of not necessarily being transient? Our computer simulations lead us to offer the following conjecture, which can in fact (tediously) be proved mathematically.

In any stable configuration, players must be aligned on a straight line in a "stripe" wrapped around the Torus. The stripes must be at least two players thick if horizontal or vertical, and at least three players thick if diagonal.[12] Any discontinuity will generate a process of erosion of the cluster that will inevitably lead to its disappearance.

Things become relatively more complicated if $\tilde{m} \neq 4$. Here as well the system could indeed converge to a stable state in which both strategies coexist, but the patterns that emerge are more difficult to describe. Relying on imagination, rather than on formalism, we could describe the stable clusters that emerge in the long run as "spots," the maximum curvature of which is entirely determined by the cut-off value.

In general, one mild property of the self-organizing pattern that emerges from the simulations is worth being noticed: stripes and spots cannot coexist, and, as the difference in terms of risk between the two equilibria becomes less important (i.e., as $\tilde{m} \to 4$), the spot pattern connects up into a pattern of stripes.

5. CONCLUDING REMARKS

This paper studies a locally interactive evolutionary model, in which players are assumed to employ aspiration-based learning rules. The simple contest we analyze is that of a one-shot pure coordination game that agents repeatedly play with their neighbors. Strategy choices are formulated in terms of simple majority rules and are updated with inertia. Under these assumptions, we investigate the limit behavior of the system describing the interaction of a finite population of players. We prove that for 2-by-2 coordination games, rules of thumb of the kind described above constitute quite a robust rationale for Nash equilibrium. We show that a straightforward extension of the result to 3-by-3 coordination games is possible only when the equilibria in pure strategies are risk-equivalent.

We then focus on the characterization of the absorbing states, and, with the help of simulation results carried out on a Cellular Automaton machine, describe the patterns that can emerge in the limit. We show that, under local interaction, the process may end up in a steady state where none of the strategies available to players has disappeared. The latter finding is in sharp contrast with what would be obtained in a model that was analogous to the one we consider in the formal setting but in which interaction is assumed to be nonlocal.

The formalization presented in this paper allows for a quite general class of economic applications, ranging from mimetic contagion in financial markets (as in Kirman [1991]), to dynamics of technological diffusion (as those studied in a nonlocal setting in Dosi and Kaniovski [1993]), or phenomena of price dispersion (as we propose in Anderlini and Ianni [1995]). In general, there seems to be no a priori reason not to extend the analysis to many economic domains in which the following key ingredients are present: existence of (positive) network externalities, complex communication structure in the population that motivates boundedly rational "imitative" behavior on the part of agents, such as aspiration-driven learning rules. As the foregoing formalization suggests, while path-dependency applies throughout, informational imperfections or the complexity of the environment fosters variety, that is, the coexistence, in equilibrium, of different strategy choices.

NOTES

1. Note that these two conditions imply that neighborhoods are overlapping: at least one of i's neighbors' neighbors is certainly i's neighbor as well.
2. Here, and throughout the paper, the symbol $\|A\|$ refers to the cardinality of the set A.
3. Whenever possible, the subscript t will be dropped for notational convenience.
4. Aspiration-driven learning is, for example, analyzed in Bendor, Mookherjee, and Ray (1991); and Binmore and Samuelson (1993).
5. This is, in essence, the reason why equilibrium-selection issues cannot be addressed in this framework.

6. The indices i and j are used either to refer to *strategies* or to *players*: this should not generate confusion, given that each player is entirely identified by (his location and) the strategy he chooses while playing G.
7. $I(\cdot)$ denotes the identity map.
8. Note that the latter case is possible *only* if $m < N$ (i.e., under the assumption of local interaction) and amounts to saying that, in a steady state, both strategies coexist.
9. Casti (1991) provides an excellent introduction to the dynamics of Cellular Automata, as well as an exhaustive bibliography. As further references, we recall the work of Toffoli and Margolus (1987); and Wuinsche and Lesser (1992).
10. Assuming a random threshold is likely to sensibly alter the analysis. "Moody" players are, for example, those considered by Binmore and Samuelson (1993).
11. Totalistic rules are those where the value of a cell at time $t + 1$ depends only on the sum of the values of the cells in the neighborhood at time t.
12. These are the same steady states that we identify in Anderlini and Ianni (1995).

REFERENCES

Anderlini, L., and A. Ianni. 1995. "Path Dependence and Learning From Neighbours." *Games and Economic Behavior*, forthcoming.
Bendor, J., D. Mookherjee, and D. Ray. 1991. "Aspiration-Based Adaptive Learning in Two Person Repeated Games." *Mimeo*, Indian Statistical Institute.
Berninghaus, K. S., and U. Schwalbe. 1992. "Learning and Adaptation Processes in Games with a Local Interaction Structure." *Mimeo*, University of Mannheim.
Binmore, K. 1987. "Modeling Rational Players, Part I." *Economics and Philosophy* 3:179–214.
Binmore, K. 1988. "Modeling Rational Players, Part II." *Economics and Philosophy* 4:9–55.
Binmore, K., and L. Samuelson. 1993. "Muddling Through: Noisy Equilibrium Selection." *Mimeo*, University College London.
Blume, L. E. 1993. "The Statistical Mechanics of Strategic Interaction." *Games and Economic Behavior* 5:387–424.
Canning, D. 1992. "Average Behavior in Learning Models." *Journal of Economic Theory* 57:442–72.
Casti, J. L. 1991. *Reality Rules*. Vol. 1. New York: Wiley.
van Damme, E. 1987. *Stability and Perfection of Nash Equilibria*. Berlin: Springer-Verlag.
Dosi, G., and Y. Kaniovski. 1993. "The Method of Generalized Urn Scheme in the Analysis of Technological and Economic Dynamics." *Mimeo*, IIASA WP-93-17.
Ellison, G. 1991. "Learning, Local Interaction, and Coordination." *Mimeo*, MIT.
Friedman, D. 1991. "Evolutionary Games in Economics." *Econometrica* 59:637–66.
Fudenberg, D., and D. Kreps. 1990. "A Theory of Learning, Experimentation, and Equilibrium in Games." *Mimeo*, MIT.
Harsanyi, J. C., and R. Selten. 1988. *A General Theory of Equilibrium Selection in Games*. Cambridge: MIT Press.
Kandori, M., G. J. Mailath, and R. Rob. 1993. "Learning, Mutation, and Long Run Equilibria in Games." *Econometrica* 61:29–56.
Kandori, M., and R. Rob. 1993. "Bandwagon Effects and Long Run Technology Choice." *Mimeo*, University of Pennsylvania.
Kirman, A. 1992. "Variety: The Coexistence of Techniques." *Revue d'Économie Industrielle* 59:65–79.
Nachbar, J. H. 1990. "Evolutionary Selection Dynamics in Games: Convergence and Limit Properties." *International Journal of Game Theory* 19:59–90.

Robert, F. 1986. *Discrete Iterations*. Berlin: Springer-Verlag.

Samuelson, L., and J. Zhang. 1992. "Evolutionary Stability in Asymmetric Games." *Journal of Economic Theory* **57**:121–39.

Toffoli, T., and N. Margolus. 1987. *Cellular Automata Machines*. Cambridge: MIT Press.

Wuinsche, A., and M. Lesser. 1992. *The Global Dynamics of Cellular Automata*. New York: Addison-Wesley.

Young, H. P. 1993. "The Evolution of Conventions." *Econometrica* **61**:57–84.

6

Evolutive vs. naive Bayesian learning

IMMANUEL M. BOMZE, JÜRGEN EICHBERGER

Abstract

Evolutive learning provides a game-theoretic interpretation of the replicator dynamics. There are two different ways to extend the discrete evolutive learning process to learning dynamics in continuous time. This paper studies the analogy between evolutive learning and the naive Bayesian learning process introduced by Eichberger et al. (1993). It introduces continuous-time versions of the naive Bayesian learning model, examines existence and uniqueness of dynamic learning trajectories, and studies their properties. We show that Bayesian learning does not change the support of the prior distribution. Furthermore, the updating process of the average probability of any pure strategy is governed by the covariance between the probability of this pure strategy and the probability of the observed strategy. For the 2×2 case, the dynamics of evolutive and naive Bayesian learning are investigated in detail. In this case, the two learning models exhibit a strikingly similar behavior over time.

1. INTRODUCTION

Evolutive learning is a game-theoretic interpretation of the replicator dynamics that plays a prominent role in biomathematical models of the evolution of two populations without self-interaction. In a bimatrix game, the present mixed strategy is updated according to the relative performance of pure strategies given the opponent's (previous) mixed strategy. In the evolutive learning model, there are two different ways to extend the discrete learning process to learning dynamics in continuous time.

In contrast to this approach, the naive Bayesian learning process, introduced by Eichberger et al. (1993), models a situation in which each player, using the previous response of the opponent, updates his belief about the opponent's choice of a mixed strategy before choosing a best reply to the updated belief. In

During the time when this research was done, Jürgen Eichberger was a member of the Department of Economics at the University of Melbourne. Both authors would like to acknowledge the support of the Department of Economics.

analogy to the evolutive learning model, there are two continuous representations of the discrete learning dynamics. This paper introduces these continuous versions of the naive Bayesian learning model. It examines existence and uniqueness of dynamic-learning trajectories and studies their properties. We show that Bayesian learning does not change the support of the prior distribution. Furthermore, the updating process of the average probability of any pure strategy is governed by the covariance between the probability of this pure strategy and the probability of the observed strategy.

For the 2×2 case, the dynamics of evolutive and naive Bayesian learning are investigated in detail. It is remarkable that, although originating from completely different approaches, both learning models exhibit a strikingly similar behavior over time. Thus, "empirically" payoff-driven learning may produce the same dynamics as Bayesian updating of beliefs in the light of the opponent's behavior combined with expected-payoff maximizing choices of strategies.

The paper is organized as follows. Section 2 reviews the evolutive learning model and its continuous-time extensions. In addition, it describes the evolutive learning dynamics of the 2×2 model for the discrete and continuous-adjustment cases. The naive Bayesian learning model is presented in section 3. Its continuous-time versions are introduced, and a proof of existence and uniqueness of learning trajectories is given. Two subsections derive general properties of the naive Bayesian learning process and study its dynamics in 2×2 games.

2. REPLICATOR DYNAMICS AS AN EVOLUTIVE LEARNING MODEL

2.1. Introduction: Discrete time dynamics

Consider a bimatrix game $\Gamma(A, B)$, where $A = [a_{ij}]$ designates the $m \times n$ payoff matrix for player 1 having m pure strategies represented by the standard basis vectors $e_1, \ldots, e_m$ in $\mathbb{R}^m$, while the $m \times n$ matrix $B = [b_{ij}]$ comprises payoffs for player 2 having n pure strategies represented by the standard basis vectors $f_1, \ldots, f_n$ in $\mathbb{R}^n$ (the d-dimensional Euclidean space is denoted by $\mathbb{R}^d$).

Consequently, a mixed strategy x of player 1 belongs to the convex hull S^m of $e_1, \ldots, e_m$, that is, S^m is the standard simplex in $\mathbb{R}^m$. Similarly, player 2 can use mixed strategies $y \in S^n$, the convex hull of $f_1, \ldots f_n$. As usual, a strategy profile $(x^*, y^*) \in S^m \times S^n$ is said to be a Nash equilibrium point if it satisfies

$$x^T A y^* \leq (x^*)^T A y^* \quad \text{and} \quad (x^*)^T B y^* \leq (x^*)^T B y^* \quad \text{for all } (x, y) \in S^m \times S^n,$$

where v^T denotes the transpose of a column vector v.

In biological models of frequency-dependent selection (an approach that is often termed "biological game theory"), the quantities above have the following meaning: consider a conflict between members of two populations without self-interaction, for example, the question of sharing the investment in offspring between males and females of a certain species. Assume that with respect to this conflict, any individual displays one out of m (resp. n) possible behavior patterns, which do not change during its whole lifetime, and which are inherited by its offspring. Then the payoff matrices comprise the scores of incremental fitness (e.g., numbers of offspring) resulting from this behavior. The mixed strategies then are simply describing the state of the population, for example, $x = [x^i] \in S^n$ consists of the relative frequencies x^i of individuals displaying e_i within population 1 (since we reserve subscripts to denote time instants or numbers, the i-th coordinate of a column vector $v \in \mathbb{R}^d$ is always denoted by a superscript, v^i, with the exception of v^T, the transpose of v). Assuming random encounters, the mean payoff for an individual displaying e_i is given by

$$(2.1) \qquad e_i^T Ay = \sum_{j=1}^{n} a_{ij} y^j,$$

if population 2 is in state $y = [y^j] \in S^n$.

Hence the average mean payoff within population 1 in state $x \in S^n$ amounts to

$$(2.2) \qquad x^T Ay = \sum_{i=1}^{m} x^i e_i^T Ay = \sum_{i=1}^{n} \sum_{j=1}^{n} a_{ij} x^i y^j.$$

The relative success of a behavior pattern e_i may now be defined by the ratio of quantities in (2.1) and (2.2). Replicator dynamics relates the time evolution of the relative frequencies x_t^i to this ratio, that is, postulates for the next generation at time $t + \Delta t$

$$(2.3a) \qquad x_{t+\Delta t}^i = x_t^i \frac{e_i^T Ay_t}{x_t^T Ay_t}, \quad 1 \le i \le m,$$

and, by analogy,

$$(2.3b) \qquad y_{t+\Delta t}^j = y_t^j \frac{x_t^T Bf_j}{x_t^T By_t}, \quad 1 \le i \le n.$$

Of course one may add a constant $C > 0$ to the quantities (2.1) and (2.2) that corresponds to some background fitness. Then (2.3) becomes

$$(2.4) \qquad x_{t+\Delta t}^i = x_t^i \frac{e_i^T Ay_t + C}{x_t^T Ay_t + C}, \quad 1 \le i \le m,$$

$$y_{t+\Delta t}^j = y_t^j \frac{x_t^T Bf_j + C}{x_t^T By_t + C}, \quad 1 \le i \le n.$$

If C is suitably large, then the denominators in (2.4) do not vanish on the state (or strategy) space $S = S^m \times S^n$ and (2.4) is a dynamical system in discrete time leaving S invariant. In terms of game theory, one may interpret (2.3) or (2.4) as follows: at each stage t a series of frequently repeated games will be performed, during which both players constantly use their mixed strategies x_t, and y_t, respectively. After this series, both empirically adjust their strategies according to the relative success for the pure strategies they employed. So both concentrate on the actual payoff outcome but refrain from calculating best answers given the opponent's behavior. Thus this setting incorporates only a low level of rationality, if any A learning process coherent with this model can therefore be termed as evolutive (or naive, empirical) learning.

Nevertheless, the dynamics (2.3) or (2.4) lead in some sense to a rationally explainable outcome.

Example 2.1 *The matching-pennies game.*

Put $A = \begin{bmatrix} 1 & 0 \\ 0 & 1 \end{bmatrix}$ and $B = \begin{bmatrix} 0 & 1 \\ 1 & 0 \end{bmatrix}$. Then (2.3) reduces to a dynamics on the unit square for $\xi = x^1$ and $\eta = y^1$, since $x^2 = 1 - \xi$ and $y^2 = 1 - \eta$:

$$\xi_{t+\Delta t} = \frac{\xi_t \eta_t}{\xi_t \eta_t + (1 - \xi_t)(1 - \eta_t)},$$

$$\eta_{t+\Delta t} = \frac{\eta_t(1 - \xi_t)}{\eta_t(1 - \xi_t) + \xi_t(1 - \eta_t)}.$$

It helps to introduce $a = \frac{1-\xi}{\xi}$ and $b = \frac{1-\eta}{\eta}$, because then $a_{t+\Delta t} = a_t b_t$ and $b_{t+\Delta t} = b_t/a_t$. Iterating these relations, we obtain an almost periodic pattern of evolution within 8 generations: starting with $a_0 = a$ and $b_0 = b$, we obtain, putting $n_k = 16^k$, for the times $t = (8k + j)\Delta t, k = 0, 1, 2, \ldots,$

j	0	1	2	3	4	5	6	7
a_t	a^{n_k}	$(ab)^{n_k}$	b^{2n_k}	$\left(\frac{b}{a}\right)^{2n_k}$	a^{-4n_k}	$(ab)^{-4n_k}$	b^{-8n_k}	$\left(\frac{a}{b}\right)^{8n_k}$
b_t	b^{n_k}	$\left(\frac{b}{a}\right)^{n_k}$	a^{-2n_k}	$(ab)^{-2n_k}$	b^{-4n_k}	$\left(\frac{a}{b}\right)^{4n_k}$	a^{8n_k}	$(ab)^{8n_k}$

Hence for all starting values (ξ_0, η_0) in the interior of the unit square, for sufficiently large t the trajectories come arbitrarily close to all of the four corners representing pure-strategy pairs in a cycling manner with the same period length $8\Delta t$. For instance, if $\xi_0 = \frac{1}{3}$, then $\xi_{12\Delta t} = (1 + 2^{-64})^{-1} \approx 1 - 10^{-19}$, while $\xi_{16\Delta t} = (1 + 2^{256})^{-1} \approx 1 - 10^{-77}$, making computer simulations *absurd. However, the time averages*

$$\bar{\xi}_t = \frac{1}{t} \sum_{s=0}^{t-1} \xi_s \quad \text{and} \quad \bar{\eta}_t = \frac{1}{t} \sum_{s=0}^{t-1} \eta_s$$

do converge to $\frac{1}{2}$ and hence $(\bar{x}_t, \bar{y}_t) \to (x^*, y^*)$, the unique Nash equilibrium point, as $t \to \infty$.

2.2. Replicator dynamics in continuous time: Two variants

A straightforward way to obtain a system of differential equations from (2.4) is to replace the finite difference $x_{t+\Delta t}^i - x_t^i$ with the time derivative $\dot{x}_t^i$, and so on. Then we arrive at

$$(2.5) \qquad \dot{x}_t^i = x_t^i \frac{e_i^T A y_t - x_t^T A y_t}{x_t^T A y_t + C}, \quad 1 \le i \le m,$$

$$\dot{y}_t^j = y_t^j \frac{x_t^T B f_j - x_t^T B y_t}{x_t^T B y_t + C}, \quad 1 \le j \le n.$$

Another approach takes into account that differentials are limits of difference quotients $[x_{t+\Delta t}^i - x_t^i]/\Delta t$ rather, and replaces these expressions with $\dot{x}_t^i$, identifying the small generation time delay Δt with $1/C$, which in turn is close to the inverse of the denominators occurring in (2.4) and (2.5) if C is large. The resulting dynamics then is

$$(2.6) \qquad \dot{x}_t^i = x_t^i [e_i^T A y_t - x_t^T A y_t], \quad 1 \le i \le m,$$

$$\dot{y}_t^j = y_t^j [x_t^T B f_j - x_t^T B y_t], \quad 1 \le j \le n.$$

One might prefer (2.5)—in particular with $C = 0$—in case of absolute payoffs or fitnesses. On the other hand, if background fitness C is large (2.6) seems to be more appropriate. Both dynamics have been studied in detail (see Maynard Smith 1982, appendix J, and Hofbauer and Sigmund 1988, chapter 27), so that we can refer to their main results for the 2×2 case of Example 2.1: under (2.5), convergence of (x_t, y_t) to the Nash equilibrium point (x^*, y^*) is guaranteed for any starting point (x_0, y_0) in the interior of S. In contrast, under (2.6) we have periodically oscillating trajectories with period τ, that is $(x_\tau, y_\tau) = (x_0, y_0)$, then

$$(2.7) \qquad \frac{1}{\tau} \int_0^\tau x_t dt = x^* \quad \frac{1}{\tau} \int_0^\tau y_t dt = y^*,$$

so that the time averages over one period are *exactly* identical with the Nash equilibrium strategy pair (x^*, y^*).

So in this case (a unique, totally mixed Nash equilibrium strategy pair), all three dynamics behave very differently in the long run but have in common that one might obtain the Nash equilibrium by inspecting the time averages of the trajectories.

In all other generic cases (one player with a dominant strategy, or three Nash equilibria, two of them consisting of pure strategies and one totally mixed equilibrium), all three dynamics (2.3), (2.5), and (2.6) have the same behavior: almost all trajectories converge to one of the pure strategy equilibria. In the three equilibria cases, the selection of one depends only on the starting values (x_0, y_0).

The proof of these assertions follows lines similar to the analysis in Schuster et al. (1981) and will be omitted here. As we shall see in section 3, the same phenomenon emerges if one considers a rational dynamical learning model introduced by Eichberger et al. (1993).

3. NAIVE BAYESIAN LEARNING DYNAMICS

3.1. Introduction: Discrete time dynamics

The following model has been introduced by Eichberger et al. (1993). For the reader's convenience we shortly recapitulate their approach, using a slightly different notation: every player has, at instant t, a certain belief about his or her opponent's use of mixed strategies, represented by probability measures, P_t on S^n for player 1 and Q_t on S^m for player 2. According to his or her belief, player 1 expects to face an average mixed strategy

$$(3.1) \qquad q_t = \int_{S^n} q \, P_t(dq)$$

and chooses a (pure) a best reply $x_t \in r_1(q_t)$. Similarly, player 2 expects

$$(3.2) \qquad p_t = \int_{S^m} p \, Q_t(dp)$$

and chooses a (pure) best reply $y_t \in r_2(p_t)$. By r_i we denote the best-reply correspondences for player i, which, generically, collapse to one point at p_t, and q_t, respectively, for most of the time. If r_i consists of more than one point we may and do leave the choice of x_t or y_t undefined. In particular, for the learning dynamics in continuous time, which we shall present in section 4, this ambiguity will play no role.

Then the game is performed: player 1 plays x_t and player 2 plays y_t. This information is used by the players to update their beliefs as follows:

$$(3.3) \qquad P_{t+\Delta t}(M) = \frac{1}{q_t^j} \int_M q^j \, P_t(dq), \quad \text{if } y_t = f_j,$$

where $M \subseteq S^n$ is a Borel set and $q_t^j = \int_{S^n} q^j P_t(dq)$ for $1 \le j \le n$. Relation (3.3) is just the conditional probability of q lying in M, given $y_t = f_j$ is

114

observed, expressed by means of Bayes's formula, interpreting $P_t \times q$ as the joint distribution of player 2's is mixed strategy q and his or her choice f_j of pure strategy. This relation can be written more compactly as

$$(3.4) \qquad P_{t+\Delta t}(M) = \left[\int_{S^n} \varphi_t(q) P_t(dq) \right]^{-1} \int_M \varphi_t(q) P_t(dq),$$
$$M \text{ a Borel subset of } S^n,$$

with

$$(3.5) \qquad \varphi_t(q) = \sum_{j=1}^{n} q^j f_j^T y_t, \quad q \in S^n.$$

Observe that φ_t depends on time t only via $y_t \in r_2(p_t)$, which in turn is ("almost" uniquely) determined by Q_t, due to (3.2).

Similarly, player 2 updates his belief

$$(3.6) \qquad Q_{t+\Delta t}(N) = \left[\int_{S^m} \psi_t(p) Q_t(dp) \right]^{-1} \int_N \psi_t(p) Q_t(dp),$$
$$N \text{ a Borel subset of } S^m,$$

with

$$(3.7) \qquad \psi_t(p) = \sum_{i=1}^{m} p^i e_i^T x_t, \quad p \in S^m.$$

Again, ψ_t depends only via P_t on t.

For any topological space X, let $\mathcal{M}(X)$ denote the space of (signed) Borel probability measures on X. Then one can view (3.4) and (3.6) as a discrete-time dynamical system operating on the product $\mathcal{M}(S^m) \times \mathcal{M}(S^n)$,

$$(3.8) \qquad \left. \begin{aligned} P_{t+\Delta t} &= \left[\int_{S^n} \Phi(Q_t) dP_t \right]^{-1} \Phi(Q_t) \cdot P_t \\ Q_{t+\Delta t} &= \left[\int_{S^m} \Psi(P_t) dQ_t \right]^{-1} \Psi(P_t) \cdot Q_t \end{aligned} \right\},$$

with $\Phi(Q_t)(q) = \varphi_t(q)$ and $\Psi(P_t)(p) = \psi_t(p)$. For any Borel-measurable function f defined on X, the expression $f \cdot P$ means the (signed) Borel measure on X given by

$$(3.9) \qquad (f \cdot P)(M) = \int_M f(q) P(dq), \quad M \text{ a Borel subset of } X.$$

Notice that the payoffs as specified in the matrices A and B do not enter the dynamics (3.8) directly, but via the best-reply correspondences r_i. Furthermore, the expression at the right-hand side of (3.8) is undefined if $q_t^j = 0$ and $y_t = f_j$

(or $p_t^i = 0$ and $x_t = e_i$), which can happen only if $q^j = 0$ holds P_t-almost surely (or if $p^i = 0$ holds Q_t-almost surely).

Example 3.1 *In example 2.1, we have*

$$
r_1(q) = \begin{cases} \{e_2\}, & \text{if } 0 \le q^1 < \frac{1}{2}, \\ S^2, & \text{if } q^1 = \frac{1}{2}, \\ \{e_1\}, & \text{if } \frac{1}{2} < q^1 \le 1, \end{cases} \quad \text{and } r_2(p) = \begin{cases} \{f_1\}, & \text{if } 0 \le p^1 < \frac{1}{2}, \\ S^2, & \text{if } p^1 = \frac{1}{2}, \\ \{f_2\}, & \text{if } \frac{1}{2} < p^1 \le 1. \end{cases}
$$

This game has only one Nash equilibrium point, namely, $\left(\left[\frac{1}{2}, \frac{1}{2}\right]^T \left[\frac{1}{2}, \frac{1}{2}\right]^T \right)$. Let us start with a priori distributions

$$
P_0 = \frac{1}{4}\delta_{f_1} + \frac{3}{4}\delta_{f_2} \quad \text{and} \quad Q_0 = \frac{1}{4}\delta_{e_1} + \frac{3}{4}\delta_{e_2}
$$

concentrated on the pure strategies of the opponent (by δ_p we always denote a point measure located p). Then $r_1(q_0) = \{e_2\}$ and $r_2(p_0) = \{f_1\}$. Therefore, according to (3.4) and (3.6), $P_{\Delta t}$ is concentrated on f_1, while $Q_{\Delta t}$ is concentrated on e_2. Consequently, $q_{\Delta t} = f_1$ and $p_{\Delta t} = e_2$. Hence $r_1(q_{\Delta t}) = \{e_1\}$ and $r_2(p_{\Delta t}) = \{f_1\}$, which leaves $Q_{2\Delta t}$ undefined. However, if we start with the a priori beliefs concentrated on $p = q = [0.9, 0.1]^T \in S^2$ and $p' = q' = [0.1, 0.9]^T \in S^2$, that is, consider

$$
P_0 = \frac{1}{4}\delta_q + \frac{3}{4}\delta_{q'} \quad \text{and} \quad Q_0 = \frac{1}{4}\delta_p + \frac{3}{4}\delta_{p'},
$$

then any difficulties are removed, and we obtain (at least via numerical evidence) for the expected strategies p_t and q_t a picture that is very similar to that in Example 2.1: p_t^1 and q_t^1, and hence (p_t, q_t), do not converge as $t \to \infty$ but cycle through the four quadrants of the unit square with increasing passage times, coming arbitrarily close to the corners as t gets large. However, the time averages

$$
\bar{p}_t = \frac{1}{t}\sum_{s=0}^{t-1} p_s \quad \text{and} \quad \bar{q}_t = \frac{1}{t}\sum_{s=0}^{t-1} q_s
$$

seem to converge to the Nash equilibrium strategy pair as $t \to \infty$ (see Table 1).

As Example 3.1 shows, for the 2×2 case the dynamics (3.8) are reduced to a system of difference equations operating on pairs of probability distributions over $p^1 \in [0, 1]$, and $q^1 \in [0, 1]$. The time evolution of these distributions is investigated in detail by Eichberger et al. (1993), who show, among other results, that the family of beta distributions is invariant under (3.8) and obtain sufficient conditions for convergence in this case. Note that beta distributions are a special case of the exponential family (with $m = n = 2$), which is shown to be time-invariant in section 3.4 below.

116

Table 1. *Evolution of expected strategies and their time averages for the matching pennies example under dynamics (3.8)*

Period t	q_t	$\bar{q}_t$	p_t	$\bar{p}_t$
1	0.30	0.30	0.30	0.30
2	0.70	0.50	0.13	0.21
3	0.87	0.62	0.30	0.24
4	0.90	0.69	0.70	0.36
5	0.87	0.73	0.87	0.46
6	0.70	0.72	0.90	0.53
7	0.30	0.66	0.90	0.59
8	0.13	0.60	0.90	0.62
9	0.10	0.54	0.87	0.65
10	0.10	0.50	0.70	0.66
11	0.10	0.46	0.30	0.62
12	0.10	0.43	0.13	0.58
13	0.10	0.41	0.10	0.55
14	0.13	0.39	0.10	0.51
15	0.30	0.38	0.10	0.49
16	0.70	0.40	0.10	0.46
17	0.87	0.43	0.10	0.44
18	0.90	0.45	0.10	0.42
19	0.90	0.48	0.10	0.41
20	0.90	0.50	0.13	0.39
21	0.90	0.52	0.30	0.39
22	0.90	0.54	0.70	0.40
23	0.90	0.55	0.87	0.42
24	0.90	0.57	0.90	0.44
25	0.90	0.58	0.90	0.46
26	0.90	0.59	0.90	0.48
27	0.87	0.60	0.90	0.49
28	0.68	0.60	0.90	0.51
29	0.28	0.59	0.90	0.52
30	0.11	0.58	0.90	0.53
31	0.09	0.56	0.90	0.55
32	0.09	0.55	0.90	0.56
33	0.08	0.53	0.90	0.57
34	0.08	0.52	0.90	0.58
35	0.08	0.51	0.87	0.58
36	0.08	0.49	0.68	0.59
37	0.08	0.48	0.28	0.58
38	0.10	0.47	0.11	0.57
39	0.10	0.46	0.09	0.55
40	0.10	0.45	0.09	0.54

3.2. Continuous time models, I.: Derivation, existence, and uniqueness

Let us proceed as in section 2 and consider two different approaches to obtaining a system of differential equations from (3.8). First replace the finite differences $P_{t+\Delta t}(M) - P_t(M)$ with the differential $\dot{P}_t(M) = \frac{d}{dt}[P_t(M)]$ and proceed analogously with Q_t to obtain

$$
\begin{aligned}
\dot{P}_t(M) &= \left[\int_{S^n} \Phi(Q_t)dP_t\right]^{-1} \left\{(\Phi(Q_t) \cdot P_t)(M)\right. \\
&\quad \left. - \left[\int_{S^n} \Phi(Q_t)dP_t\right] P_t(M)\right\} \\
\dot{Q}_t(N) &= \left[\int_{S^m} \Psi(P_t)dQ_t\right]^{-1} \left\{(\Psi(P_t) \cdot Q_t)(N)\right. \\
&\quad \left. - \left[\int_{S^m} \Psi(P_t)dQ_t\right] Q_t(N)\right\}
\end{aligned}
$$

(3.10)

This of course models an instantaneous adaptation process and requires players with quick reaction (which in the real world would be the case, e.g., in a stock-exchange scenario, perhaps with computer-supported sell/buy decisions).

The second standard procedure to obtain a continuous-time dynamical system from the discrete-time system (3.8) replaces the difference quotient $\frac{1}{\Delta t}[P_{t+\Delta t}(M) - P_t(M)]$ with $\dot{P}_t(M)$ and identifies $\frac{1}{\Delta t}$ with the denominators in (3.8). This yields

$$
\begin{aligned}
\dot{P}_t(M) &= (\Phi(Q_t) \cdot P_t)(M) - \left[\int_{S^n} \Phi(Q_t)dP_t\right] P_t(M) \\
\dot{Q}_t(N) &= (\Psi(P_t) \cdot Q_t)(N) - \left[\int_{S^m} \Psi(P_t)dQ_t\right] Q_t(N)
\end{aligned}
$$

(3.11)

For ease of exposition, let us start investigating the latter dynamics. If the limiting process yielding the derivatives in (3.11) is performed uniformly with respect to all Borel sets $M \subseteq S^n$ and $N \subseteq S^m$, then one obtains a coupled system of differential equations on $\mathcal{M}(S^n) \times \mathcal{M}(S^m)$, equipped with the (product) variational norm on this product space:

$$
\begin{aligned}
\dot{P}_t &= \left[\Phi(Q_t) - \int_{S^n} \Phi(Q_t)dP_t\right] \cdot P_t \\
\dot{Q}_t &= \left[\Psi(P_t) - \int_{S^m} \Psi(P_t)dQ_t\right] \cdot Q_t
\end{aligned}
$$

(3.12)

Here $\dot{P}_t$ denotes the signed measure on S^n satisfying $||\frac{1}{\Delta t}[P_{t+\Delta t} - P_t] - \dot{P}_t|| \to 0$ as $\Delta t \searrow 0$, where we denote, for a signed Borel measure R on S^n, by

$$||R|| = \sup\{R(M) - R(S^n \setminus M) : M \text{ a Borel subset of } S^n\}$$

the variational norm of μ.

Of course, simple continuity conditions on Φ and Ψ can be specified that in general guarantee existence and uniqueness of trajectories satisfying (3.12); for example, see Lang (1972). One problem here is that our functions $\Phi(Q_t) = \varphi_t$ and $\Psi(P_t) = \psi_t$ may be discontinuous in interesting cases, since they depend on x_t and y_t, which in turn may jump from one vertex of S^m or S^n to another, according to the best reply correspondences r_i. See (3.4) through (3.9).

However, in generic cases these jumps will occur only at a discrete set of time points. During the intervals between these instances, $\Phi(Q_t)(q) = \varphi_t(q)$ coincides with a coordinate projection $\pi_j(q) = q^j$ that is constant over time within the interval. The only influence of Q_t is the determination of the switching points via p_t and y_t.

Therefore we obtain the following result:

Theorem 1 *Consider two a priori beliefs (P_0, Q_0) such that*

(a) $r_1(q) = \{e_i\}$ for all q in a neighborhood of p_0, or
(b) $r_2(p) = \{f_j\}$ for all p in a neighborhood of q_0.

Then generically there is a unique trajectory $(P_t, Q_t) \in \mathcal{M}(S^n) \times \mathcal{M}(S^m)$ as $t \geq 0$, starting at (P_0, Q_0) and satisfying (3.11) up to a discrete set of time points t, which is continuous and piecewise differentiable with respect to the variational norm on $\mathcal{M}(S^n) \times \mathcal{M}(S^m)$. Furthermore the measures P_t and Q_t are probability distributions that have the same null sets (and hence the same supports) as P_0 and Q_0, respectively.

PROOF: In the proof we make use of the property that "most of the time" the system (3.11) "discouples" in the sense that there is only a feedback from P_t to $\dot{P}_t$. To be more specific, observe that under the assumptions of the propositions, we have $\varphi_t(q) = q^j$ and $\psi_t(p) = p^i$ if $t \in [0, \tau)$, where $\tau \geq 0$ is the first switching time point, for example, because $r_1(q_\tau)$ consists of more than one point and therefore x_τ, and hence φ_τ is undefined. Generically, $r_2(p_\tau)$ consists still of one point, so that, generically, $\varphi_t(q) = q^j$ holds even for all $t \in [0, \tau + \varepsilon)$ where $\varepsilon > 0$ may be small. Relation (3.11) then decomposes into two autonomous systems, namely,

$$(3.13) \qquad \dot{P}_t = [\pi_j - q_t^j] \cdot P_t, \quad 0 \leq t < \tau + \varepsilon,$$

with starting distribution P_0 and

$$(3.14) \qquad \dot{Q}_t = [\pi_j - q_t^j] \cdot Q_t, \quad 0 \leq t < \tau,$$

with starting distributions Q_0. Then, as in the proof of Theorem 8 in Bomze (1991), we see that there is a unique trajectory (P_t) for $t \in [0, \tau + \varepsilon)$ starting in P_0 and satisfying (3.13), and also a unique trajectory (Q_t) for $t \in [0, \tau)$ starting in Q_0 and satisfying (3.14). Furthermore, all Q_t are absolutely continuous with respect to Q_0, with Radon/Nikodym density

$$(3.15) \quad \frac{dQ_t}{dQ_0}(p) = C_t \exp\left(\int_0^t \psi_s(p)ds\right) = C_t \exp(tp^i), \quad p \in S^m,$$

if $0 < t < \tau$. This follows as in Lemma 2 of Bomze (1991). Here $C_t = \exp(-\int_0^t p_s^i ds)$ is a normalizing constant satisfying

$$(3.16) \quad C_t^{-1} = \int_{S^m} \exp(tp^i)Q_0(dp).$$

From (3.16) we conclude $e^{-t} \leq C_t \leq 1$ and similarly also from (3.15)

$$(3.17) \quad e^{-t} \leq \frac{dQ_t}{dQ_0}(p) \leq e^t \quad \text{for all } t \in [0, \tau) \text{ and all } p \in S^m.$$

We now define Q_τ by specifying its Radon/Nikodym density with respect to Q_0 through continuation of the formula (3.15), that is, replacing $t < \tau$ with τ. As is easily seen, this yields a probability distribution on S^m satisfying $\|Q_t - Q_\tau\| \to 0$ as $t \nearrow \tau$. Moreover, for all $t \in [0, \tau]$, the distributions Q_t and Q_0 have the same null sets, since their respective Radon/Nikodym densities are bounded from above and also bounded away from zero. It remains to continue the dynamics for Q_t as $t > \tau$, starting from Q_τ. To this end we employ the fact that P_t is defined for all $t \in [0, \tau + \varepsilon)$ yielding expected strategies q_t for t in this interval. Generically, $r_1(q_t)$ will again consist of a single strategy if $\tau < t < \tau + \varepsilon$ (take a smaller $\varepsilon > 0$ if necessary), e_k, say. The corresponding dynamics for Q_t then reads

$$(3.18) \quad \dot{Q}_t = [\pi_k - p_t^k] \cdot Q_t, \quad \tau \leq t < \tau + \varepsilon,$$

with starting distribution Q_τ. Now suppose that $r_2(p_{\tau+\varepsilon})$ consists of more than one point. Then we repeat the same procedure for P_t, observing that also

$$(3.19) \quad \frac{dP_t}{dP_0}(q) = D_t \exp\left(\int_0^t \varphi_s(q)ds\right) = D_t \exp(tq^j), \quad q \in S^n,$$

if $0 < t < \tau + \varepsilon$ holds. Since both Radon/Nikodym densities are positive and integrate to 1 under Q_0 or P_0, respectively, we obtain thus in a unique way trajectories of probability distributions on S^m or S^n, respectively, which have the same null sets as the starting distributions. □

Now let us turn toward the dynamics (3.10). As above, we obtain the following dynamical system on $\mathcal{M}(S^n) \times \mathcal{M}(S^m)$:

$$
(3.20) \qquad \left. \begin{array}{l} \dot{P}_t = \Lambda(P_t, Q_t) \cdot P_t \\[12pt] \dot{Q}_t = \Xi(P_t, Q_t) \cdot Q_t \end{array} \right\}
$$

with

$$
(3.21) \qquad \left. \begin{array}{l} \Lambda(P, Q) = \left[\int_{S^n} \Phi(Q)dP \right]^{-1} \left\{ \Phi(Q) - \left[\int_{S^n} \Phi(Q)dP \right] \right\} \\[16pt] \Xi(P, Q) = \left[\int_{S^m} \Psi(P)dQ \right]^{-1} \left\{ \Psi(P) - \left[\int_{S^m} \Psi(P)dQ \right] \right\} \end{array} \right\}
$$

which involve Φ and Ψ, so that the same proviso as for (3.11) applies also to (3.21).

With moderately higher technical effort, one may prove a counterpart of Theorem 1 for this system under the following assumptions: if the starting beliefs (P_0, Q_0) satisfy (a) and (b) from Theorem 1 and if furthermore

(3.22) q_0 and p_0 belong to the relative interior of S^n and S^m, respectively,

then existence and uniqueness of continuous and piecewise differentiable trajectories under (3.21) is generically guaranteed. Again, P_t and Q_t have the same null sets as P_0 and Q_0, respectively.

3.3. Continuous time models, II.: General properties

One of the most important general properties of both dynamical models (3.11) and (3.21) is the "naivete of learning": most of the time the two equations govern independently of each other the evolution of the beliefs P_t, Q_t, until a switching point, say determined by P_t, is reached where the dynamics for Q_t is changed drastically (the same of course applies to the dynamics for P_t).

Theorem 2 *Let (P_t, Q_t) be a time-continuous trajectory satisfying (3.11) or (3.21). Then*

(a) *the supports of P_t and Q_t remain constant over time*

$$\operatorname{supp} P_t = \operatorname{supp} P_0 \text{ and } \operatorname{supp} Q_t = \operatorname{supp} Q_0 \quad \text{for all } t \geq 0.$$

(b) *the beliefs are fixed; that is, $(P_t, Q_t) = (P_0, Q_0)$ holds for all $t \geq 0$ if and only if*

(3.23) $P_0(\{q \in S^n : q^j = q_0^j\}) = Q_0(\{p \in S^m : p^i = p_0^i\}) = 1$

holds, provided i and j satisfy $r_1(q_0) = (e_i)$ and $r_2(p_0) = \{f_j\}$ (in case of [3.21] we additionally have to assume that [3.22] is satisfied).

121

PROOF: Assertion (a) follows from the fact that P_0 and Q_0 have the same null sets as P_t and Q_t, respectively. To show (b), we employ also Theorem 1 (and its counterpart): indeed, under the condition (3.23) we conclude for all $t \geq 0$ that $\Phi(Q_t)(q) = q^j = q_0^j$ for P_t-almost all $q \in S^n$ and similarly that $\Psi(P_t)(p) = p^i = p_0^i$ for Q_t-almost all $p \in S^m$. It follows that the right-hand sides of (3.11) and (3.21) vanish. The reverse argument is then also evident. $\square$

In particular, the condition in Theorem 2(b) is satisfied if the starting beliefs $P_0 = \delta_q$ and $Q_0 = \delta_p$ are concentrated at the strategies q and p for some $p \in S^m$ and $q \in S^n$. The next simple case is $P_t \neq \delta_q$, but $Q_t = \delta_p = Q_0$, due to Theorem 2(a). This case is treated, among others, in Theorem 3.

Theorem 3 *If $r_2(p) = \{f_j\}$ for Q_0-almost all $p \in S^m$, then under (3.11)*

(a) *P_t converges weakly to the set of all probability distributions that are concentrated on the set*

$$M = \{q \in \text{ supp } P_0 : q^j = \gamma\} \subseteq S^n,$$

where $\gamma = max \{q^j : q \in \text{ supp } P_0\}$ is the maximal value of q^j in the support of P_0.

(b) *if furthermore, $r_1(q) = \{e_i\}$ holds in a neighborhood of M, then Q_t converges weakly to the set of all probability measures concentrated at the maximizers of p^i on the support of Q_0.*

(c) *if above maximizers are unique, say p and q, then we have $P_t \to \delta_q$ as well as $Q_t \to \delta_p$ as $t \to \infty$.*

PROOF: (a) The condition on $r_2(p)$ means that $y_t = f_j$ holds for all t. Now let $F_t(q) = F(q) - \int_{[0,1]} F\, dP_t$ where $F(q) = q^j$ does not depend on Q_t (and also not on P_t). Then evolution of P_t according to (3.11) is given by

$$\dot{P}_t = F_t \cdot P_t.$$

Hence the assertion follows as in Example 1 of Bomze (1990).

(b) A continuity argument and (a) show that $q_t = \int_{S^n} q\, P_t(dq)$ belongs to the neighborhood addressed in the formulation of the theorem, at least for all sufficiently large $t \geq 0$. Hence also $x_t \in r_1(q_t) = e_i$ for all those t, so that we also get from (3.11)

$$\dot{Q}_t = G_t \cdot Q_t$$

with $G_t(p) = G(p) - \int_{S^n} G\, dQ_t$ where $G(p) = p^i$. Hence the result. Assertion (c) now follows easily. $\square$

Of course, similar assertions hold if $r_1(q) = \{e_i\}$ for P_0-almost all p.

The above theorem also has an important consequence in case player 2 has a dominating strategy, f_j, say. If both P_0 and Q_0 have full support, we (generically) obtain $(P_t, Q_t) \to (\delta_q, \delta_p)$ as $t \to \infty$, where $(p, q) = (e_i, f_j)$ is the (generically) unique Nash equilibrium point; compare case (iii) of Proposition 2.1 in Eichberger et al. (1993).

3.4. Continuous time models, III.: Evolution of Radon/Nikodym densities and expected strategies

For the discussion to follow, it is useful to decompose the positive time axis $(0, \infty)$ (up to a discrete number of switching points) into mutually disjoint intervals $J_{l,v} = (s_{l,v}, t_{l,v})$ and $I_{k,\mu} = (u_{k,\mu}, v_{k,\mu})$ which are characterized by the relations

$$(3.24) \qquad \left. \begin{array}{rl} y_s = f_l & \text{if} \quad s \in J_{l,v}, \\[2mm] x_s = e_k & \text{if} \quad s \in I_{k,\mu}. \end{array} \right\}$$

If we are interested in evolution up to time t, only those subscripts $v, \mu \in \{0, 1, 2, \ldots\}$ are essential which do not exceed

$$(3.25) \qquad \left. \begin{array}{l} v_l(t) = \max\{v : t_{l,v} < t\}, \\[2mm] \mu_k(t) = \max\{\mu : s_{k,\mu} < t\}. \end{array} \right\}$$

Observe that in the above notation it does not matter whether or not $v_l(t) \nearrow \infty$ or $\mu_k(t) \nearrow \infty$ as $t \nearrow \infty$.

Now from (3.15), (3.19), and the chain rule for Radon/Nikodym densities, we obtain the general formulae for the evolution under (3.11):

$$(3.26) \qquad \left. \begin{array}{l} \dfrac{dP_t}{dP_0}(q) = D_t \exp \left\{ \displaystyle\sum_{i \neq j} q^l \sum_{v=1}^{v_l(t)} (t_{l,v} - s_{l,v}) \right. \\[6mm] \qquad\qquad \left. + q^j \left[\displaystyle\sum_{v=1}^{v_j(t)-1} (t_{j,v} - s_{j,v}) + (t - s_{j,v_j(t)}) \right] \right\}, \quad \text{if } y_t = f_j, \\[10mm] \dfrac{dQ_t}{dQ_0}(p) = C_t \exp \left\{ \displaystyle\sum_{k \neq i} p^k \sum_{\mu=1}^{\mu_k(t)} (v_{k,\mu} - u_{k,\mu}) \right. \\[6mm] \qquad\qquad \left. + p^i \left[\displaystyle\sum_{\mu=1}^{\mu_i(t)-1} (v_{i,\mu} - u_{j,\mu}) + (t - u_{i,\mu_i(t)}) \right] \right\}, \quad \text{if } x_t = e_i, \end{array} \right\}$$

123

where C_t and D_t are normalizing constants as in (3.15) and (3.19). Observe that all time increments occurring in each line of (3.26) sum up to total time t.

Under evolution according to (3.21), the relative Radon/Nikodym densities satisfy

$$
\begin{aligned}
\frac{d P_t}{d P_0}(q) &= e^{-t} \exp\left(\sum_{j=1}^{m} \alpha_t^j q^j \right), \\
\frac{d Q_t}{d Q_0}(p) &= e^{-t} \exp\left(\sum_{i=1}^{n} \beta_t^i p^i \right),
\end{aligned}
$$

(3.27)

where the coefficients α_t^j and β_t^i are determined as follows: suppose again that $y_t = f_j$ and $x_t = e_i$. Then

(3.28)
$$
\alpha_t^l = \sum_{v=1}^{v_l(t)} \int_{s_{l,v}}^{t_{l,v}} (q_s^l)^{-1} ds \quad \text{if } l \neq j, \quad \text{and}
$$

$$
\alpha_t^j = \sum_{v=1}^{v_j(t)-1} \int_{s_{j,v}}^{t_{j,v}} (q_s^j)^{-1} ds + \int_{s_{j,v_j(t)}}^{t} (q_s^j)^{-1} ds,
$$

as well as

(3.29)
$$
\beta_t^k = \sum_{\mu=1}^{\mu_k(t)} \int_{u_{k,\mu}}^{v_{k,\mu}} (p_s^k)^{-1} ds \quad \text{if } k \neq i, \quad \text{and}
$$

$$
\beta_t^i = \sum_{\mu=1}^{\mu_i(t)-1} \int_{u_{i,\mu}}^{v_{i,\mu}} (p_s^i)^{-1} ds + \int_{u_{i,\mu_i(t)}}^{t} (p_s^i)^{-1} ds.
$$

Theorem 4 *Under both dynamics (3.11) and (3.21), the exponential family with densities $C \exp(\sum_{j=1}^{n} \alpha^j q^j)$ and $D \exp(\sum_{j=1}^{m} \beta^i p^i)$ is time-invariant.*

PROOF: Follows from (3.26) and (3.27). □

An important issue is the dynamical behavior of the expected strategies p_t or q_t of the respective opponent. Here the following covariance formula is essential, which has a counterpart in Price's evolution equation (Bomze 1990, p. 80) – another analogy to evolutive learning.

Theorem 5

(a) *Under (3.11), we obtain*

(3.30)
$$
\begin{aligned}
\dot{p}_t^k &= \mathrm{Cov}_{Q_t}(p^i, p^k) \quad \text{if } x_t = e_i \\
\dot{q}_t^l &= \mathrm{Cov}_{P_t}(q^j, q^l) \quad \text{if } y_t = f_j
\end{aligned}.
$$

124

(b) Under (3.21), we have

$$(3.31) \qquad \left. \begin{array}{l} \dot{p}_t^k = (p_t^i)^{-1} \mathrm{Cov}_{Q_t}(p^i, p^k) \quad \text{if } x_t = e_i \\ \\ \dot{q}_t^l = (q_t^j)^{-1} \mathrm{Cov}_{P_t}(q^j, q^l) \quad \text{if } y_t = f_j \end{array} \right\}.$$

PROOF: This follows in a straightforward manner from the calculation

$$(3.32) \qquad \dot{p}_t = \int_{S^m} p \dot{Q}_t(dp) = \int_{S^m} (p^i - p_t^i) p Q_t(dp)$$

and similarly for q_t. The proof for (b) is analogous. $\qquad \square$

As a consequence, we note that under both dynamics, the expected probability p_t^i that the opponent uses e_i increases if $x_t = e_i$ (similarly with q_r^j), since p_t^i is (positively proportional to) $\mathrm{Var}_{Q_t}(p^i) \geq 0$. This elementary relation should be satisfied in all reasonable learning processes.

Moreover, the coupling between P_t and Q_t in both dynamics is done via the expected mixed strategies q_t and p_t. Hence it is not surprising that it is impossible to have nonconstant beliefs centered at time-constant expectations:

Theorem 6 *Assume that $r_1(q_0) = \{e_i\}$ and that $r_2(p_0) = \{f_j\}$. Then (p_t, q_t) remain fixed under (3.11) or (3.21) if and only if (P_t, Q_t) remain fixed (again one has to assume additionally (3.22) in the latter case).*

PROOF: Sufficiency is obvious. To prove necessity, observe that the expected strategies can remain constant over time (i.e., $p_t = p_0$ and $q_t = q_0$ for all $t \geq 0$) only if $\mathrm{Var}_{Q_t}(p^i) = \mathrm{Var}_{P_t}(q^j) = 0$, which means that condition (3.23) is satisfied. Hence the result follows from Theorem 2(b). $\qquad \square$

The above theorem exhibits a further analogy to the evolutive learning models presented in section 2: here and there any Nash equilibrium point gives rise to a fixed point under the dynamics (2.5), (2.6), (3.30), and (3.31), but not vice versa. Indeed, take any strategy profile $(p_0, q_0) \in S^m \times S^n$ such that $r^2(p_0) = \{f_j\}$ and $r_1(q_0) = \{e_i\}$, and consider a pair of beliefs (P_0, Q_0) satisfying $\int_{S^m} p Q_0(dp) = p_0$ and $\int_{S^n} q P_0(dq) = q_0$, as well as condition (3.23). For instance put $P_0 = \delta_{q_0}$ and $Q_0 = \delta_{p_0}$. Then (P_0, Q_0) constitute a fixed point under (3.11) and (3.21), and hence (p_0, q_0) are fixed under (3.30) and (3.31). It has recently been shown that *asymptotically stable* fixed points of (3.30) and (3.31) are Nash equilibria, provided the prior distributions have full support (Eichberger 1995).

3.5. Continuous time models, IV.: The 2×2 case

Now let us consider the 2×2 ease and, for ease of notation, put again $\xi_t = p_t^1$ and $\eta_t = q_t^1$. From the definitions in (3.24), we see that the switching times $t_{l,\nu}$ are, if they exist, determined by the relation $\eta_t = \bar{\eta}$ for $t = t_{l,\nu}$ and $l \in \{1, 2\}$; similarly the condition $\xi_t = \bar{\xi}$ for $t = \nu_{k,\mu}$ determines the switching points $\nu_{k,\mu}$ and $k \in \{1, 2\}$. Here $([\bar{\xi}, 1 - \bar{\xi}]^T, [\bar{\eta}, 1 - \bar{\eta}]^T)$ is a mixed-strategy Nash equilibrium point. Note from $p^2 = 1 - p^1$ and $q^2 = 1 - q^1$ that $\mathrm{Cov}_{Q_t}(p^1, p^2) = -\mathrm{Var}_{Q_t}(p^1)$ and also $\mathrm{Cov}_{P_t}(q^1, q^2) = -\mathrm{Var}_{P_t}(q^1)$.

First we deal with the simpler dynamics (3.11). Theorem 5(a) entails $\dot{\xi}_t = \pm \mathrm{Var}_{Q_t}(p^1)$ and $\dot{\eta}_t = \pm \mathrm{Var}_{P_t}(q^1)$, where the signs of the right-hand sides depend on x_t and y_t, respectively. From Theorem 6 it follows that ξ_t and η_t can remain constant over time only if Q_t is concentrated on p_t and P_t is concentrated on q_t. However, if ξ_t and η_t are allowed to move, then their dynamics are quite simple: if $y_t = f_1$, then ξ_t increases, and so on.

Also the dynamics of the relative Radon/Nikodym densities simplifies considerably: from (3.26) it follows that

$$
(3.34) \quad
\left.
\begin{aligned}
\frac{dP_t}{dP_0}(q) &= d_t \exp[(\gamma_t + \sigma t)q^1] \quad \text{with} \quad \sigma = 1, \quad \text{if } y_t = f_1, \\
&\text{and} \quad \sigma = -1, \quad \text{if } y_t = f_2, \\
\frac{dQ_t}{dQ_0}(p) &= c_t \exp[(\zeta_t + \rho t)p^1] \quad \text{with} \quad \rho = 1, \quad \text{if } x_t = e_1, \\
&\text{and} \quad \rho = -1, \quad \text{if } x_t = e_2,
\end{aligned}
\right\}
$$

where γ_t and ζ_t are constant in the intervals $J_{l,\nu}$ and $I_{k,\mu}$, respectively, while c_t, d_t depend continuously on t.

Example 3.2 *Reconsider Example 3.1. This game has only one Nash equilibrium point, namely, $\left(\left[\frac{1}{2}, \frac{1}{2}\right]\right)^T, \left(\left[\frac{1}{2}, \frac{1}{2}\right]^T\right)$. Imagine two players whose starting beliefs are concentrated on the opponent's pure strategies. From Theorem 2(a) it follows that for all $t \geq 0$, their beliefs are of the form*

$$
P_t = \eta_t \delta_{f_1} + (1 - \eta_t)\delta_{f_2} \quad \text{and} \quad Q_t = \xi_t \delta_{e_1} + (1 - \xi_t)\delta_{e_2}.
$$

Then $\mathrm{Var}_{P_t}(q^1) = \eta_t(1 - \eta_t)$ and similarly $\mathrm{Var}_{Q_t}(p^1) = \xi_t(1 - \xi_t)$. This and Theorem 5(a) allow us to solve the differential equations for ξ_t and η_t explicitly. For $y_t = f_1$, we obtain logistic growth

$$
\xi_t = (1 + a \exp(-t))^{-1}.
$$

126

The integration constant a has to be fitted to the starting condition. To be more specific, the following conditions are equivalent:

$$\xi_\tau = \bar{\xi},$$

(3.35)
$$\left.\begin{array}{c} \tau = \log[a/(\bar{\xi}^{-1} - 1)], \\ a = e^\tau[\bar{\xi}^{-1} - 1], \end{array}\right\}$$

while for $y_t = f_2$ we obtain the same result after time reversal:

$$\xi_t = (1 + a \exp(t))^{-1}$$

with the equivalences

$$\xi_\tau = \bar{\xi},$$

(3.36)
$$\left.\begin{array}{c} \tau = \log[(\bar{\xi}^{-1} - 1)/a], \\ a = e^{-\tau}[\bar{\xi}^{-1} - 1]. \end{array}\right\}$$

Of course, the very same relations hold for η_t. We denote the corresponding integration constant by b. Let us start with $0 < \xi_0 < \frac{1}{2}$ and $0 < \eta_0 < \frac{1}{2}$. Then for small $t > 0$, we have $y_t = f_1$ so that η_t will grow logistically, while $x_t = e_2$ and ξ_t will decrease. Hence the first switching point occurs if $\eta_t = \frac{1}{2}$, that is, is of type $v_{i,\mu}$. After that point, ξ_t starts to increase and eventually (at $t = t_{j,\nu}$) reaches the value $\frac{1}{2}$. To be more specific, one obtains from (3.34) and (3.35) the following sequence of switching points and integration constants: $u_{2,0} = s_{1,0} = 0$ with $a_0 = \xi_0^{-1} - 1$ and $b_0 = \eta_0^{-1} - 1$. For notational convenience only, we will in the sequel use these constants instead of the starting conditions. Next we obtain $v_{2,0} = u_{1,0} = \log a_0$, and from continuity of η_t and $u_{1,0}$ we obtain $b_1 = a_0^2 b_0$. The next switching point is $t_{1,0} = s_{2,0} = \log(a_0^2 b_0)$, whence $a_1 = a_0^{-3} b_0^{-2}$. Thereafter we get $v_{1,0} = u_{2,1} = \log(a_0^3 b_0^2)$ and $b_2 = a_0^{-4} b_0^{-3}$, and then $t_{2,0} = s_{1,1} = \log(a_0^4 b_0^3)$, whence $a_2 = a_0^5 b_0^4$ results. Finally we arrive at $v_{2,1} = u_{1,1} = \log(a_0^5 b_0^4)$, at which point $\eta_{v_{2,1}} = \eta_{v_{2,0}}$, so that we have periodic motion with period $v_{2,1} - v_{2,0} = 4 \log(a_0 b_0)$, which is exactly the double of $u_{2,1} - v_{2,0} = 2 \log(a_0 b_0)$.

Now we establish the analogous result for the time averages: from the differential equations for η_t, we derive the identities

$$\log(\eta_{v_{2,1}}/\eta_{u_{2,1}}) = \int_{v_{2,1}}^{u_{2,1}} \frac{\dot{\eta}_t}{\eta_t} dt = \int_{v_{2,1}}^{u_{2,1}} (\eta_t - 1) dt = \int_{v_{2,1}}^{u_{2,1}} \eta_t dt - (v_{2,1} - u_{2,1}),$$

$$\log(\eta_{v_{1,0}}/\eta_{u_{1,0}}) = \int_{v_{1,0}}^{u_{1,0}} \frac{\dot{\eta}_t}{\eta_t} dt = \int_{v_{1,0}}^{u_{1,0}} (1 - \eta_t) dt$$

$$= -\int_{u_{1,0}}^{v_{1,0}} \eta_t dt + (v_{1,0} - u_{1,0}).$$

127

Using $v_{1,0} = u_{2,1}$ and $v_{2,0} = u_{1,0}$, we thus arrive at

$$\int_{v_{2,0}}^{v_{2,1}} \eta_t dt = v_{2,1} - v_{2,0} + \log(\eta_{v_{2,0}}^2/\eta_{v_{1,0}}^2) = 2\log(a_0 b_0),$$

and hence the time average of q_t over a period is exactly the Nash equilibrium strategy $[\frac{1}{2}, \frac{1}{2}]^T$. Since the same arguments hold for p_t, we are faced with the exact counterpart of the situation depicted in section 2.2; see (2.7).

Now let us investigate the dynamics (3.21): as above we obtain

$$(3.37) \qquad \dot{\xi}_t = \begin{cases} (\xi_t)^{-1} \mathrm{Var}_{Q_t}(p^1), & \text{if } y_t = f_1, \\ -(1-\xi_t)^{-1} \mathrm{Var}_{Q_t}(p^1), & \text{if } y_t = f_2, \end{cases}$$

and

$$(3.38) \qquad \dot{\eta}_t = \begin{cases} (\eta_t)^{-1} \mathrm{Var}_{P_t}(q^1), & \text{if } x_t = e_1, \\ -(1-\eta_t)^{-1} \mathrm{Var}_{P_t}(q^1), & \text{if } x_t = e_2, \end{cases}$$

Example 3.3 *We continue with the game above, but under the evolution described by (3.21). The differential equations are*

$$\dot{\xi}_t = \begin{cases} 1-\xi_t, & \text{if } y_t = f_1, \\ -\xi_t, & \text{if } y_t = f_2, \end{cases}$$

and similar for η_t. We can integrate these and obtain

$$\xi_t = \begin{cases} 1-ae^{-t}, & \text{if } y_t = f_1, \\ ae^{-t}, & \text{if } y_t = f_2. \end{cases}$$

Again, a is the integration constant satisfying one of the following equivalent conditions:

$$(3.39) \qquad \left. \begin{array}{l} \xi_\tau = \bar{\xi}, \\ \tau = \log[a/(1-\bar{\xi})], \\ a = e^\tau (1-\bar{\xi}), \end{array} \right\}$$

if $y_t = f_1$, or

$$(3.40) \qquad \left. \begin{array}{l} \xi_\tau = \bar{\xi}, \\ \tau = \log[a/\bar{\xi}], \\ a = e^\tau \bar{\xi}, \end{array} \right\}$$

if $y_t = f_2$. Proceeding as in the example above, but using (3.38) and (3.39) instead, and denoting by b the integration constant for η_t, we get $u_{2,0} = s_{1,0} = 0$ with $a_0 = 1 - \xi_0$ and $b_0 = \eta_0$. Again, we will use these constants instead of *the starting conditions. Next we obtain $v_{2,0} = u_{1,0} = \log(2a_0)$ and from continuity of η_t at $u_{1,0}$ we obtain $b_1 = 2a_0 - b_0$. The next switching point is $t_{1,0} = s_{2,0} = \log(4a_0 - 2b_0)$ whence $a_1 = 3a_0 - 2b_0$. Thereafter we get $v_{1,0} = u_{2,1} = \log(6a_0 - 4b_0)$ and $b_2 = 4a_0 - 3b_0$, and then $t_{2,0} = s_{1,1} = \log(8a_0 - 6b_0)$ whence $a_2 = 5a_0 - 5b_0$ results. Finally we arrive at $v_{2,1} = u_{1,1} = \log(10a_0 - 8b_0)$ with*

$$\eta_{v_{2,1}} = \frac{4a_0 - 3b_0}{10a_0 - 8b_0} = f(c_0) < c_0 = \frac{b_0}{2a_0} = \eta_{v_{2,0}} < \frac{1}{2},$$

since $f(c) = (2-3c)/(5-8c)$ is strictly concave for $0 \le c \le \frac{1}{2}$, and $f(0) = \frac{2}{5}$ while $f(\frac{1}{2}) = \frac{1}{2}$ so that $\frac{1}{2}$ is the only fixed point of f on the considered interval. Similarly, we conclude

$$\xi_{t_{1,1}} = \frac{7 - 3c_0}{12 - 5c_0} > \frac{3 - c_0}{4 - c_0} = \xi_{t_{1,0}} > \frac{1}{2},$$

so that we arrive at $\xi_t \to \frac{1}{2}$ and $\eta_t \to \frac{1}{2}$ in an oscillating manner as $t \nearrow \infty$. Hence the expected strategies converge to the Nash equilibrium strategies under (3.21) as under the evolutive learning dynamics (2.5).

4. CONCLUSION

In this paper, we presented some remarkably parallel features between evolutive learning models that incorporate only a low level of individual rationality, and the naive Bayesian learning process introduced by Eichberger et al. (1993). For three qualitatively different dynamics, we obtained in the 2×2 case of the matching-pennies game essentially the same long-run results: divergence in the discrete-time model, cycling under one continuous-time dynamics, and convergence under the other. All three (or six) models have in common that one can recover the unique Nash equilibrium point consisting of mixed strategies by looking at the time averages of trajectories. Although there is numerical evidence that this strong analogy breaks down in higher-dimensional cases, for example, with the Shapley counterexample for fictitious-play dynamics (see, e.g., Rosenmüller[1971]), recent findings exhibit a similar phenomenon for many games, the so-called AR-BR principle (Gaunersdorfer and Hofbauer 1995; Hofbauer 1995).

REFERENCES

Bomze, I. M. 1990. "Dynamical aspects of evolutionary stability." *Monatsh. Math.* **110**:189–206.
Bomze, I. M. 1990. "Cross entropy minimization in uninvadable states of complex populations." *J. Math. Biol.* **30**:73–87.
Eichberger, J. 1995. "Bayesian learning in repeated normal form games." *Games and Economic Behaviour* **11**, 2 (Nov. 1995).

Eichberger, J., H. Haller, and F. Milne. 1993. "Naive Bayesian learning in 2×2 matrix games." *J. of Economic Behaviour and Organization* **22**:69–90.

Gaunersdorfer, A. and J. Hofbauer. 1995. "Fictitious play, Shapley polygons and the replicator equation." *Games and Economic Behaviour* **11**, 2 (Nov. 1995).

Hofbauer, J. (1995). "Rational versus evolutionary dynamics for games." In R. Cressman and G. Hines, eds. *Dynamic evolutionary game theory in biology and economics.* Conference report, Wilfrid Laurier Univ. and the Univ. of Guelph.

Hofbauer, J. and K. Sigmund. 1988. *The theory of evolution and dynamical systems.* Cambridge Univ. Press.

Maynard Smith, J. 1982. *Evolution and the theory of games.* Cambridge Univ. Press.

Lang, S. 1972. *Differential manifolds.* Reading, Mass.: Addison-Wesley.

Rosenmüller, J. 1971. "Über Periodizitätseigenschaften spieltheoretischer Lernprozesse." *Z. Wahrscheinlichkeitsheo. verw. Geb* . **17**:259–308.

Schuster, P., K. Sigmund, J. Hofbauer, and R. Wolff. 1981. "Selfregulation of behaviour in animal societies II: Games between two populations without selfinteraction." *Biological Cybernetics* **40**:9–15.

7

Learning and mixed-strategy equilibria in evolutionary games

VINCENT P. CRAWFORD

Abstract

This paper considers whether Maynard Smith's concept of an evolutionarily stable strategy, or "ESS," can be used to predict long-run strategy frequencies in large populations whose members are randomly paired to play a game, and who adjust their strategies over time according to sensible learning rules. The existing results linking the ESS to stable equilibrium population strategy frequencies when strategies are inherited do not apply to learning, even when each individual always adjusts its strategy in the direction of increased fitness, because the inherited-strategies stability results depend on aggregating across individuals, and this is not possible for learning. The stability of learning must therefore be analyzed for the entire system of individuals' strategy adjustments. The interactions between individuals' adjustments prove to be generically destabilizing at mixed-strategy equilibria, which are saddlepoints of the learning dynamics. Using the inherited-strategies dynamics to describe learning implicitly restricts the system to the stable manifold whose trajectories approach the saddlepoint, masking its instability. Thus, allowing for the interactions between individuals' strategy adjustments extends the widely recognized instability of mixed-strategy equilibria in multi-species inherited-strategies models to single-species (or multi-species) learning models.

1. INTRODUCTION

The concept of an evolutionarily stable strategy, or "ESS," was introduced by Maynard Smith & Price (1973) and Maynard Smith (1974) to describe the effects of selection for more successful strategies in environments where an individual's expected rate of reproduction, or fitness, is jointly determined by its own and other individuals' strategies. In the model originally analyzed by Maynard Smith and most often studied by subsequent writers, individuals are selected at random from a population and matched, in pairs, to play a symmetric two-person game. The individuals in the population are identical except for their strategies; these are inherited and fixed for life. The population is large enough that the differences between the expected strategy frequencies faced by different individuals are negligible, even though individuals are never paired with themselves and generally play different strategies. Individuals reproduce asexually and breed true, passing on their strategies unchanged to their offspring. Finally, an individual's current fitness is jointly determined by its strategy and the strategy of the individual with which it is currently paired, as summarized by the payoff matrix of the game.[1]

Maynard Smith defined an ESS for the model just described as a mixed strategy that, if played by all members of a *monomorphic* population, has strictly higher fitness than any mutant strategy that enters the population with sufficiently low frequency. (The definition is the same for *polymorphic* populations, in which individuals play different strategies in equilibrium, with the qualification that mutants must then have lower fitness than the population, on average.) The intuition for Maynard Smith's definition is that, if the members of a population all play an ESS, mutants that enter the population with low frequency will reproduce more slowly than individuals who play the ESS. The mutants' relative population frequencies will therefore approach zero over time, restoring the population strategy frequencies of the ESS. Computing an ESS should then allow the analyst to infer the possible long-run values of these frequencies from the payoff matrix alone, an important simplification.

This intuition was first formalized, for continuous-time versions of the population dynamics, by Taylor & Jonker (1978), Zeeman (1979), and Hines (1980a, b).[2] Taylor & Jonker (1978) and Zeeman (1979) studied symmetric two-person finite matrix games, requiring individuals to play pure strategies, and considered large polymorphic populations, in which different strategies may persist in equilibrium. They showed that, for generic payoffs, a vector of population strategy frequencies that (when treated as a mixed strategy) satisfies the ESS condition with arbitrary mixed strategies allowed as mutations is a locally asymptotically stable (henceforth "stable") equilibrium of the population dynamics.[3] Taylor & Jonker (1978) gave an example to show that the converse is not true in general, so that Maynard Smith's definition is overly restrictive when individuals are required to play pure strategies: Games whose players have more than two pure strategies can have equilibria whose strategy frequencies do not satisfy the ESS condition, such that any small group of pure-strategy mutants with higher average fitness than the population also has individual fitness differences that alter its strategy frequencies over time in such a way that the population strategy frequencies return to the equilibrium.

Hines (1980 a, b) (see also Cressman & Hines, 1984; Hines, 1987, section 4; Maynard Smith, 1974, 1982, chapter 2 and appendix D; Zeeman, 1979, 1981) identified a closer link between the ESS condition and stability, showing that when individuals in a large polymorphic or monomorphic population are allowed to play mixed strategies, the ESS condition is generically necessary as well as sufficient for stability of the population strategy frequencies. A population whose strategy frequencies violate the ESS condition is vulnerable to any mixed-strategy mutant with higher fitness, because the strategy frequencies of such mutants (unlike those of groups of pure-strategy mutants) do not evolve when there are fitness differences between their constituent pure strategies. This result unifies the treatment of pure and mixed strategies on the individual level and shows how to use the concept of evolutionary stability, originally formulated for monomorphic populations, to characterize the long-run effects of strategy selection in polymorphic populations of mixed strategies.

It is important in what follows to note that, because the results just summarized relate *relative* strategy frequencies and fitnesses, they apply equally well to growing, fixed, or shrinking populations. Also, because a large population playing a strategy that does not maximize fitness against itself is clearly vulnerable to a low-frequency mutation that does maximize fitness against that strategy, an ESS must be a symmetric Nash equilibrium of the game played by matched pairs.[4] The stability arguments of Taylor & Jonker, Zeeman, and Hines can therefore be viewed as an alternative justification for this standard game-theoretic characterization of behavior as well as for the ESS.

In recent years, the idea of evolutionary stability has been applied extensively in biology, and its usefulness has been found surprisingly robust to deviations from the original population model (see for example Hines, 1987, sections 4–6). Perhaps encouraged by this robustness, a number of biologists and social scientists have suggested using the ESS to explain behavior in human or animal populations in which inheritance of strategies is supplemented or supplanted by learning (Axelrod, 1984; Harley, 1981; Hines, 1987; Hines & Bishop, 1983; Houston & Sumida, 1987; Maynard Smith, 1982; chapters 5, 13; Sugden, 1986; Zeeman, 1979, 1981). Such applications rest implicitly on a dynamic justification like that developed for inherited strategies by Taylor & Jonker, Zeeman, and Hines. Learning plainly fits the inherited-strategies model if it proceeds purely by imitation, with members of successive generations choosing strategies, once and for all, in numbers proportional to the payoffs of earlier adherents of those strategies. But this rules our individual strategy adjustment, an essential feature of learning. This paper considers whether it is possible to construct a sensible justification for using the ESS to describe the consequences of learning that involves individual strategy adjustment.

The issues raised by individual strategy adjustment stand out most clearly when it is the only source of of change in population strategy frequencies. From now on, I shall use the term "learning" in this special sense, further restricting attention for simplicity to fixed populations.

There are important similarities between the inherited-strategies dynamics and sensible learning dynamics, because if each individual in a large population adjusts in the direction of increased payoffs, the population strategy frequencies also move in that direction. For this reason, it is often assumed (see, for example, Axelrod, 1984; Sugden, 1986; Taylor & Jonker, 1978: 146, 153; Zeeman, 1981: 251) that the inherited-strategies justification for the ESS extends to learning. It is shown here, however, that mixed-strategy equilibria (but not pure-strategy equilibria, in general) are generically unstable for sensible specifications of the learning dynamics. It follows that if the learning dynamics converge, they must converge to a configuration in which individuals play only pure strategies.

Learning and inherited strategies can have different implications for stability because the results of Taylor & Jonker, Zeeman, and Hines depend on aggregating across individuals; aggregation is justified in large populations for inherited strategies, but not, in general, for learning. Learning must therefore be analyzed at the individual level, taking into account the interactions between individuals' strategy adjustments. These prove to be generically destabilizing at mixed-strategy equilibria, which are saddlepoint equilibria of the learning dynamics. Using the aggregate inherited-strategies dynamics to describe learning in effect restricts these dynamics to the stable manifold whose trajectories approach the saddlepoint, masking this instability.

The rest of the paper is organized as follows. Section 2 defines the ESS for large populations and reviews its relationship to the symmetric Nach equilibria of the game played by matched pairs, and to the stable equilibrium population strategy frequencies of the inherited-strategies dynamics. Section 3 compares the inherited-strategies dynamics with sensible learning dynamics, showing that the latter do not allow aggregation, even in large populations. Section 4 shows that mixed-strategy equilibria are generically unstable for the learning dynamics introduced in section 3. The stability analysis is carried out explicitly only for the "Hawk–Dove" example of Maynard Smith & Parker (1976) and Maynard Smith (1982); section 5 discusses the straightforward extension to more general symmetric two-person finite matrix games. Section 5 also discusses other extensions of the analysis and related work.

2. INHERITED STRATEGIES

This section defines the ESS for large populations and discusses its relationship to the symmetric Nash equilibria of the game played by matched pairs and to the stable equilibria of the inherited-strategies dynamics.

Consider a large population of identical individuals, repeatedly and anonymously paired at random to play a symmetric two-person finite matrix game. Recall that an ESS is a mixed strategy that, if initially played by all members of the population, has higher expected payoff than any mutant mixed strategy that enters the population with low frequency. Let $\mathbf{q}$ and $\mathbf{s}$ be vectors that give the expected frequencies with which the pure strategies are played in the population, and let $E(\mathbf{q}|\mathbf{s})$ be the expected payoff of the mixed strategy $\mathbf{q}$ when the expected population frequencies are given by $\mathbf{s}$. (With random pairing in a large population, it makes no difference whether the distribution of strategies that yields these frequencies is monomorphic or polymorphic.) An ESS can then be formally defined as a mixed strategy $\mathbf{s}$ such that for any $\mathbf{q} \neq \mathbf{s}$ and any sufficiently small $\varepsilon > 0$,

$$(1) \qquad E[\mathbf{s}|(1 - \varepsilon)\mathbf{s} + \varepsilon\mathbf{q}] > E[\mathbf{q}|(1 - \varepsilon)\mathbf{s} + \varepsilon\mathbf{q}].$$

Using the linearity in probabilities of expected payoffs reduces eqn (1)[5] to

$$(2) \qquad (1 - \varepsilon)E(\mathbf{s}|\mathbf{s}) + \varepsilon E(\mathbf{s}|\mathbf{q}) > (1 - \varepsilon)E(\mathbf{q}|\mathbf{s}) + \varepsilon E(\mathbf{q}|\mathbf{q}).$$

This inequality holds for all small $\varepsilon > 0$ if and only if, for all $\mathbf{q}$,

$$(3) \qquad E(\mathbf{s}|\mathbf{s}) \geq E(\mathbf{q}|\mathbf{s}),$$

and

$$(4) \qquad E(\mathbf{s}|\mathbf{q}) > E(\mathbf{q}|\mathbf{q}) \quad \text{whenever} \quad E(\mathbf{s}|\mathbf{s}) \geq E(\mathbf{q}|\mathbf{s}),$$

Inequality (3) simply requires $(\mathbf{s}, \mathbf{s})$ to be a symmetric Nash equilibrium in the game played by matched pairs; eqn (4) is a further implication of evolutionary stability, discussed in Maynard Smith (1982, chapter 2) and elsewhere.

The Hawk–Dove game can be used to illustrate this definition and its relationship to stability of the population strategy frequencies for the inherited-strategies dynamics. The Hawk–Dove game has payoff matrix

$$
\begin{array}{c c c}
 & H & D \\
H & A & b \\
D & c & d
\end{array}
$$

In this matrix, H and D stand for Hawk and Dove, and only the Row player's payoffs are shown; the Column player's payoffs can be deduced by symmetry.

I begin by considering monomorphic populations. Let z denote each individual's equilibrium probability of playing H (so that $s_1 \equiv z$ and $s_2 \equiv 1 - z$ in terms of the more general notation used above to define the ESS). Ignoring borderline cases for simplicity, eqns (3) and (4) imply that $z = 1$ is an ESS if and only if $a > c$, $z = 0$ is an ESS if and only if $d > b$, and $z^* = (b - d)/(b - d + c - a)$ is an ESS if and only if $c > a$ and $b > d$ (in which case $0 < z^* < 1$). The first two of these conclusions follow immediately from eqn (3). To verify the third, note that playing z^* yields $(bc - ad)/(b - d + c - a)$ against the population frequency z^*, as does any alternative mixed strategy q; thus z^* satisfies eqn (3). To satisfy eqn (4), z^* must yield a higher expected payoff against $\mathbf{q}$ than $\mathbf{q}$ does against itself, so that

$$(5) \qquad z^*[\mathbf{q}a + (1 - \mathbf{q})b] + (1 - z^*)[\mathbf{q}c + (1 - \mathbf{q})d]$$
$$> \mathbf{q}[\mathbf{q}a(1 - \mathbf{q})b] + (1 - \mathbf{q})[\mathbf{q}c + (1 - \mathbf{q})d]$$

or, equivalently,

$$(6) \qquad (z^* - \mathbf{q})\{[\mathbf{q}a + (1 - \mathbf{q})b] - [\mathbf{q}c + (1 - \mathbf{q})d]\} > 0.$$

It is easy to verify eqn (6) from the parameter restrictions $c > a$ and $b > d$.

The inherited-strategies population dynamics for this model are easiest to describe if it is assumed, following Taylor & Jonker (1978) and Zeeman (1979), that the population is polymorphic and each individual plays a pure strategy. Then the state of a large population can be summarized by its strategy frequencies, and because all individuals who play a given strategy have the same fitness, the expected proportional rate of growth of a strategy's population frequency equals the current difference between its fitness and the population frequency-weighted average fitness of all pure strategies. Let $\bar{z}$ denote the expected population frequency of H. Then $\dot{\bar{z}}$, the time rate of change of $\bar{z}$, equals $\bar{z}$ times the fitness of H minus the population average fitness when its expected frequency is $\bar{z}$:

$$(7) \qquad \dot{\bar{z}} \equiv \bar{z}\langle[\bar{z}a + (1 - \bar{z})b] - \{\bar{z}[\bar{z}a + (1 - \bar{z})b]$$
$$+ (1 - \bar{z})[\bar{z}c + (1 - \bar{z})d]\}\rangle$$
$$\equiv \bar{z}(1 - \bar{z})\{[\bar{z}a + (1 - \bar{z})b] - [\bar{z}c + (1 - \bar{z})d]\}.$$

The assumption that the population is large underlies eqn (7) in three ways: it justifies treating the population frequency as a continuous variable, it justifies using the same value of this frequency to calculate different individuals' fitnesses, and it justifies, via the law of large numbers, identifying the total realized rate of growth of the individuals playing each pure strategy and the fitness that equals its mathematical expectation. These observations can be used to show that eqn (7) holds exactly in infinite populations, and that it continues to hold, approximately, in sufficiently large finite populations.

The differential eqn (7) has three equilibria: $\bar{z} = 0, \bar{z} = 1$, and $\bar{z} = \bar{z}^* \equiv (b - d)/(b - d + c - a)$. Taylor & Jonker (1978) showed that, aside from borderline cases, each of these equilibria is stable if and only if the payoff parameters are such that it is an ESS. When $\bar{z} = \bar{z}^*$, for instance, the term in square brackets in the second line of eqn (7) equals zero, and

$$(8) \qquad \partial\dot{\bar{z}}/\partial\bar{z}|_{\bar{z}=\bar{z}^*} = \bar{z}^*(1 - \bar{z}^*)(a - b - c + d)$$
$$= (c - a)(d - b)/(b - d + c - a);$$

thus, given that $0 < \bar{z}^* < 1$, the stability condition $\partial\dot{\bar{z}}/\partial\bar{z}|_{\bar{z}=\bar{z}^*} < 0$ is equivalent to the parameter restrictions $c > a$ and $b > d$. As noted above, Taylor &

Jonker, Zeeman and Hines extended this generic equivalence between the ESS condition and stability to all symmetric two-person finite matrix games in which individuals may play mixed strategies.

3. LEARNING AND AGGREGATION

This section introduces sensible specifications of the learning process and shows that they do not allow the use of aggregate relationships like eqn (7) to describe the dynamics of strategy frequencies in infinite populations. My argument assumes a specific individual adjustment process for concreteness, but it will be clear that its conclusion does not depend on the details of the process.

The use of aggregate dynamics like eqn (7) to describe learning is often justified informally by noting that, if each individual in an infinite population adjusts its strategy in the direction of increased payoffs, the population strategy frequencies also adjust in the direction, as eqn (7) requires. Justifying eqn (7), however, also requires an aggregation argument. To see when aggregation is possible, imagine that each individual chooses a mixed strategy, adjusting it over time in response to the differences between the current payoffs of its pure strategies. Let s^i_j and s^i, respectively, denote individual i's probability of playing its jth pure strategy and i's vector of mixed-strategy probabilities. Let $\bar{s}$ denote the vector of expected population strategy frequencies, and let e_j denote the jth pure strategy, expressed as a mixed-strategy probability vector with a one in the jth place and zeroes elsewhere. Otherwise maintaining the notation and assumptions of section 2, and ignoring boundary problems, which are not germane to the point made here, assume that, for each i and j, individual i adjusts its jth mixed-strategy probability according to

$$(9) \qquad \dot{s}^i_j \equiv s^i_j [E(e_j | \bar{s}) - E(s^i | \bar{s})].$$

The differential equation system defined by eqn (9) sets the vector of proportional rates of change of each individual's mixed-strategy probabilities equal to the vector of partial derivatives of the individual's expected payoff with respect to those probabilities, computed taking into account the linearity in probabilities of expected payoffs and that the probabilities must be rescaled proportionally so they will continue to sum to one. (Summing eqn (9) over j and recalling that $E(s^i | \bar{s}) \equiv \sum_j s^i_j E(e_j | \bar{s})$ shows that $\sum_j \dot{s}^i_j \equiv 0$ for each i.)

Thus eqn (9) defines a gradient process for each individual, adjusting its mixed strategy in the current direction of greatest payoff increase in the simplex of feasible probability vectors. These gradient processes are plausible descriptions of "atheoretical" learning, in which individuals who lack either the sophistication or the knowledge of the structure of their environment to draw complex inferences from the information they receive discover how their

137

strategies influence their payoffs by experimenting occasionally with small adjustments, favoring those that yield the largest payoff improvements.

The gradient learning model defined by eqn (9) is also the closest individual-adjustment analog, in large populations, of the inherited-strategies dynamics studied by Taylor & Jonker, Zeeman, and Hines. Summing eqn (9) over i would immediately yield those dynamics if the $E(\mathbf{s}^i | \bar{\mathbf{s}})$ term on its right-hand side were replaced by $E(\bar{\mathbf{s}} | \bar{\mathbf{s}})$. But this would make $\dot{s}^i_j / s^i_j$ independent of $\mathbf{s}^i$, and eqn (9) could then no longer ensure, even in a large population, that each individual's mixed-strategy probabilities remain in the feasible simplex. Thus, the presence of the $E(\mathbf{s}^i | \bar{\mathbf{s}})$ term on the right-hand side of eqn (9) reflects an inherent difference between learning and inherited strategies.

The difference has the important consequence that, even in large populations, individuals' learning processes cannot be aggregated to give $\dot{\bar{\mathbf{s}}}$ as a function of $\bar{\mathbf{s}}$ alone. To see this, return to the Hawk–Dove game, letting z^i denote i's mixed-strategy probability of playing H and again letting $\bar{z}$ denote the expected population frequency of H. Equation (9) specializes to

$$
\begin{aligned}
(10) \quad \dot{z}^i &\equiv z^i \langle [\bar{z}a + (1 - \bar{z})b] - \{z^i [\bar{z}a + (1 - \bar{z})b] \\
&\quad + (1 - z^i)[\bar{z}c + (1 - \bar{z})d]\} \rangle \\
&\equiv z^i (1 - z^i) \{ [\bar{z}a + (1 - \bar{z})b] - [\bar{z}c + (1 - \bar{z})d] \}.
\end{aligned}
$$

Equation (10) does not yield an aggregate relationship between $\dot{\bar{z}}$ and $\bar{z}$, because its right-hand side is non-linear in z^i: In general, $\dot{\bar{z}}$ depends on the individual z^i, not just on $\bar{z}$. The impossibility of aggregation is easiest to see in a finite population with N members, where

$$
(11) \qquad \bar{z} \equiv \sum_{i=1}^{N} z^i / N
$$

and

$$
\begin{aligned}
(12) \quad \dot{\bar{z}} &\equiv \sum_{i=1}^{N} \dot{z}^i / N \\
&\equiv \{ [\bar{z}a + (1 - \bar{z})b] - [\bar{z}c + (1 - \bar{z})d] \} \sum_{i=1}^{N} z^i (1 - z^i) / N \\
&\equiv \{ [\bar{z}a + (1 - \bar{z})b] - [\bar{z}c + (1 - \bar{z})d] \} \left[\bar{z} - \sum_{i=1}^{N} (z^i)^2 / N \right],
\end{aligned}
$$

which relates $\dot{\bar{z}}$ to the individual z^i as well as $\bar{z}$. But even in a large population, the aggregation required for a learning interpretation of eqn (7) is justified only when each individual plays the same mixed strategy throughout, so that $z^i \equiv \bar{z}$ for all i.[6]

4. STABILITY

This section studies the stability of the learning dynamics introduced in section 3, showing that individuals' strategies are generically unstable at mixed-strategy equilibria. (The implications of this result for the stability of population strategy frequencies are considered in section 5.)

Because individuals' learning processes cannot be aggregated, the analysis must be carried out for the entire system of their strategy adjustments. To give a clear account of the issues that arise in specifying this system, and to describe the interactions between individuals' adjustments precisely, I work with finite populations, otherwise maintaining the assumptions and notation of sections 2 and 3. The finite-population analysis reveals what happens in large populations by passing to the limit and shows why the normal practice of doing this at the start of the analysis is misleading in this case. It is assumed throughout that the population still has enough members to justify identifying realized population strategy frequencies and their expectations.

To adapt Maynard Smith's ESS definition to finite populations, assume a fixed population of size N and write the expected payoff of individual i playing mixed strategy $\mathbf{q}$ as $E(\mathbf{q}|\bar{\mathbf{s}}^i)$, where

$$(13) \qquad \bar{\mathbf{s}}^i \equiv \sum_{\substack{j=1 \\ j \neq i}}^{N} \mathbf{s}^j / (N - 1)$$

gives the expected population strategy frequencies, *excluding* individual i. This generalizes the large-population definition given in section 2 to environments where excluding an individual's strategy from the population has a non-negligible effect on its strategy frequencies.

As before, an ESS, if played by all members of the population, must have an expected payoff at least as high as any mutant strategy that enters the population with sufficiently low frequency. The formal definition is derived, following Schaffer (1988) and Maynard Smith (1988), by taking "low frequency" to mean $1/N$, assuming that the single mutant switches strategies from the ESS (so that the population remains fixed at size N), and comparing the payoffs of the mutant and the individuals who continue to play the ESS. An ESS is now a mixed strategy $\mathbf{s}$ such that, for any $\mathbf{q} \neq \mathbf{s}$,

$$(14)[7] \qquad E\left(\mathbf{s} \left| \frac{N-2}{N-1}\mathbf{s} + \frac{1}{N-1}\mathbf{q} \right.\right) \geq E(\mathbf{q}|\mathbf{s});$$

inequality (14) is the finite-population analog of inequality (3). [Schaffer, 1988, supplemented inequality (14) with a "stability" condition that plays the role of inequality (4), showing that the ESS for the Hawk–Dove game implied by inequality (14) also satisfied this supplementary condition. The points made here, however, rely only on inequality (14).]

A useful alternative definition of the ESS for finite populations, also due to Schaffer, requires a mixed strategy **s** to solve

$$(15) \qquad \max_q E(\mathbf{q}|\mathbf{s}) - E\left(\mathbf{s}\middle| \frac{N-2}{N-1}\mathbf{s} + \frac{1}{N-1}\mathbf{q}\right),$$

where the maximization is taken over all mixed strategies **q**. This is plainly equivalent to requiring inequality (14), because inequality (14) holds with equality if $\mathbf{q} = \mathbf{s}$. Two important conclusions for finite populations follow from this equivalence. First, although any ESS converges to a symmetric Nash equilibrium as the population grows, an ESS is no longer a symmetric Nash equilibrium in general. Second, the implication of Nash equilibrium that the pure strategies in the support of a mixed ESS all yield equal expected payoffs against the expected population strategy frequencies no longer holds exactly. Instead, any mutant playing a mixture of strategies in the support of a mixed-strategy ESS has the same expected payoff as an individual playing the ESS, taking into account the fact that the individuals playing the ESS (unlike the mutant) are matched with the mutant with probability $1/(N-1)$. This follows from the linearity of the objective function in eqn (15) and the fact that its maximized value is zero.

I now argue that the equilibria (stable or unstable) of sensible learning dynamics in finite populations with sufficiently many members must be located at symmetric Nash equilibria. It follows that ESS's are equilibria of these dynamics only in large populations, in general, and that a learning justification for the ESS depends on establishing the stability of symmetric Nash equilibria in large populations.

Reiterating the second conclusion about the finite-population ESS from above, if **s** is a mixed-strategy ESS, it must satisfy

$$(16) \qquad E(\mathbf{e}_j|\mathbf{s}) = E\left(\mathbf{s}\middle| \frac{N-2}{N-1}\mathbf{s} + \frac{1}{N-1}\mathbf{e}_j\right)$$

for all pure strategies j in its support. [Requiring eqn (16) for the extreme points $\mathbf{e}_j$ of the support of **s** is a proxy, given the linearity in **q** of

$$E(\mathbf{q}|\mathbf{s}) - E\left(\mathbf{s}\middle| \frac{N-2}{N-1}\mathbf{s} + \frac{1}{N-1}\mathbf{q}\right),$$

for requiring eqn (16) for all **q** that have the same support as **s**.] For individuals' adjustment processes to reach equilibrium at **s**, they would have to respond [generalizing the expressions in eqn (16) to polymorphic populations] to differences between the $E(\mathbf{e}_j|\bar{\mathbf{s}}^i)$ and the

$$E\left(\mathbf{s}^i \left| \frac{N-2}{N-1}\bar{\mathbf{s}}^i + \frac{1}{N-1}\mathbf{e}_j\right.\right).$$

Rescaling problems (which are nontrivial but not insurmountable) aside, there seems to be no sensible way to justify this kind of response when individuals experiment independently. If individuals' experiments are rare, simultaneous experiments by matched individuals are rare indeed. Individual i's payoff experience, whether experimenting or not, is therefore generated against a population whose expected strategy frequencies are approximately $\bar{\mathbf{s}}^i$. Thus, independent experiments cannot provide the information individuals would need to respond to differences between the $E(\mathbf{e}_j|\bar{\mathbf{s}}^i)$ and the

$$E\left(\mathbf{s}^i \left| \frac{N-2}{N-1}\bar{\mathbf{s}}^i + \frac{1}{N-1}\mathbf{e}_j\right.\right).$$

This argument suggests that, at least when the population has many members, the finite-population analog of eqn (9) is a sensible model of the learning process. This model sets, for each i and j,

(17) $$\dot{\mathbf{s}}^i_j \equiv \mathbf{s}^i_j[E(\mathbf{e}_j|\bar{\mathbf{s}}^i) - E(\mathbf{s}^i|\bar{\mathbf{s}}^i)].$$

The system defined by eqn (17) is in equilibrium at symmetric Nash equilibria, but not, in general, at ESS's in finite populations. I now consider whether eqn (17) yields a learning justification for the symmetric Nash equilibrium and the ESS in sufficiently large finite populations.

The analysis is carried out explicitly only for the Hawk–Dove game; extensions to more general games are considered in section 5. Let

(18) $$\bar{z}^i \equiv \sum_{\substack{j=1 \\ j\neq i}}^{N} z^i/(N-1)$$

and specialize eqn (17) to

(19) $$\begin{aligned}
\dot{z}^i &\equiv z^i\langle[\bar{z}^i a + (1-\bar{z}^i)b] \\
&\quad - \{z^i[\bar{z}^i a + (1-\bar{z}^i)b] + (1-z^i)[\bar{z}^i c + (1-\bar{z}^i)d]\}\rangle \\
&\equiv z^i(1-z^i)\{[\bar{z}^i a + (1-\bar{z}^i)b] - [\bar{z}^i c + (1-\bar{z}^i)d]\} \\
&\equiv z^i(1-z^i)(\bar{z}^i)\alpha + \beta), \quad i = 1, \dots, N,
\end{aligned}$$

where $\alpha \equiv a - b - c + d$ and $\beta \equiv b - d$.

The differential equation system (19) is the finite-population analog of eqn (10). Like eqn (10), it allows aggregation only if individuals always play the same mixed strategies, so that z^i, and therefore $\bar{z}^i$, is independent of i. The stability analysis must therefore be carried out for the entire system.

Assume that $c > a$ and $b > d$, so that $\alpha < 0$ and $\beta > 0$, and $z^* = (b-d)/(b-d \div c-a) = -\beta/\alpha$ is the only interior equilibrium of eqn (19) and the only symmetric Nash equilibrium of the game. Partially differentiating eqn (19) and evaluating the results at $z^i = -\beta/\alpha$ for all i reveals that, for all i,

$$(20) \qquad \partial \dot{z}^i / \partial z^i = 0$$

and, for all i and all $j \neq i$,

$$(21) \qquad \partial \dot{z}^i / \partial z^j = (-\beta/\alpha)(1 + \beta/\alpha)\alpha/(N-1)$$
$$= -\beta(1 + \beta/\alpha)/(N-1) < 0.$$

The matrix of the locally linearized version of eqn (19) thus has zeroes along its main diagonal and a constant, $\kappa \equiv -\beta(1 + \beta/\alpha)(N-1)$, everywhere else.

Because the trace of a matrix equals the sum of its eigenvalues, in this case zero, at least generically one or more of the eigenvalues must have a strictly positive real part, making the system unstable. For the Hawk–Dove game, it can be shown that the eigenvalues equal $(N-1)\kappa \equiv \beta(1 + \beta/\alpha) < 0$ and $-\kappa \equiv \beta(1 + \beta/\alpha)/(N-1) > 0$, the latter having multiplicity $N-1$ (as is necessary for the eigenvalues to sum to the trace of the matrix in this case), so that the system is always unstable. The proof, due to Dennis Smallwood (personal communication), is as follows:

Letting A denote the matrix of the locally linearized system determined by eqns (20) and (21), the eigenvalues, denoted λ, and the associated eigenvectors, denoted $\mathbf{x}$, can be obtained by solving $\mathbf{Ax} = \lambda\mathbf{x}, \mathbf{x} \neq \mathbf{0}$, where $\mathbf{0}$ denotes a vector of zeroes conformable to $\mathbf{x}$. Writing out $\mathbf{Ax} = \lambda\mathbf{x}$, given that $\mathbf{A}$ has zeroes on the main diagonal and κ's everywhere else, yields:

$$(22) \qquad \kappa x_2 + \kappa x_3 + \cdots + \kappa x_{n-1} + \kappa x_n = \lambda x_1$$
$$\kappa x_1 + \kappa x_2 + \kappa x_3 + \cdots + \kappa x_{n-1} + \kappa x_n = \lambda x_2$$
$$\cdots$$
$$\cdots$$
$$\cdots$$
$$\kappa x_1 + \kappa x_2 + \kappa x_3 + \cdots + \cdots + \kappa x_n = \lambda x_{n-1}$$
$$\kappa x_1 + \kappa x_2 + \kappa x_3 + \cdots + \kappa x_{n-1} + \cdots = \lambda x_n.$$

Summing the equations in (22) yields

$$(23) \qquad \kappa(N-1)\sum_{i=1}^{n}\mathbf{x}_i = \lambda\sum_{i=1}^{n}\mathbf{x}_i.$$

Because κ and $(N-1)$ differ from zero, this has only one solution for λ when $\Sigma_{i=1}^{N}\mathbf{x}_i \neq 0$, namely $\lambda^1 \equiv \kappa(N-1)$; the associated eigenvector can be taken to be $\mathbf{x}^1 \equiv (1,\dots,1)$. Other solutions are possible when $\Sigma_{i=1}^{N}\mathbf{x}_i = 0$; it is easy to verify that the $\mathbf{x}^j \equiv (1,0,\dots,0,-1,0,\dots,0)$, where -1 is the jth component $j = 2,\dots,N$, are also eigenvectors, each with associated eigenvalue $\lambda^j \equiv -\kappa$. These N linearly independent eigenvectors and the associated eigenvalues, with multiplicity as indicated above, constitute a complete set.

Thus, all of the eigenvalues of the locally linearized system are real (as indicated by the symmetry of $\mathbf{A}$), and $N-1$ of them are strictly positive. The true, non-linear system is therefore unstable, like the linear system. The system drifts away from equilibrium approximately at an exponential rate proportional to $1/N$, hence the drift is very slow when N is large. The system is unstable, however, for any finite population.[8]

Only if $z^i \equiv \bar{z}$ throughout for all i can eqn (19) be aggregated to obtain eqn (7), and the inherited-strategies analysis be used to show that the learning dynamics are stable. The eigenvalue of the locally linearized version of eqn (7) equals $(N-1)\kappa \equiv -\beta(1+\beta/\alpha)$, the only stable eigenvalue of the locally linearized learning dynamics. Thus, using the inherited-strategies dynamics to analyze the stability of learning in effect restricts the system to its stable manifold, masking its instability.

5. CONCLUSION

This section discusses extensions of the analysis and related work. The instability result established here is mathematically closely related to the multi-species inherited-strategies results of Eshel & Akin (1983), Hines (1981) and Maynard Smith (1982, appendix J), and to the instability results established for learning processes like those considered here by Crawford (1974, 1985) in a closely related game-theoretic framework.[9] The main contribution of the present paper is to show that the already widely recognized instability of mixed-strategy equilibria for multi-species inherited-strategies dynamics extends to sensible single-species (or multi-species) learning dynamics, once the finiteness of the population and the impossibility of aggregating individuals' learning processes are recognized.[10]

The generic instability of mixed-strategy equilibria for learning dynamics extends to symmetric (or asymmetric) finite matrix games with any number of pure strategies. For any such game, the matrix of the locally linearized learning dynamics has zeroes along the main diagonal, so that, for generic payoffs, there is at least one eigenvalue with strictly positive real part. These zeroes arise because each individual's expected payoff is linear in its own mixed strategy, so that the first-order equalities required for equilibrium in individual adjustment processes like eqn (17) immediately imply that the second-order effects measured by the partial derivatives in eqn (20) equal zero. This is true of mixed-strategy equilibria in general, and their instability is therefore a robust result. In particular, individual adjustment processes with equilibria at the ESS in finite populations, as discussed in section 4, would also be unstable.

As noted in Crawford (1974, 1985) and Hines (1981), the proof given here, which ignores boundary problems, extends, generically, to equilibria in which some pure strategies are played with zero probability, as long as two or more are played with positive probability. Then, the pure strategies that have zero probability can simply be eliminated from the game and the stability analysis of section 4 carried out for the reduced game. Except in borderline cases, section 4's argument is unaffected by this reduction.

Given the general instability of mixed-strategy equilibria, pure-strategy configurations are the only possible stable outcomes of the learning dynamics. My instability arguments clearly do not apply to equilibria of these dynamics at which all individuals play pure strategies. In fact, it is not hard to see that any pure-strategy Nash equilibrium that satisfies inequality (3) with strict inequality must be a locally stable equilibrium of learning dynamics like those studied here, and analysis of simple examples suggests that pure-strategy polymorphisms may also be locally stable. [The results of Hines (1980b) and Zeeman (1981) for the inherited-strategies dynamics are inconclusive on this point for the learning dynamics, because they depend on using the large-population assumption to justify aggregation, eliminating the interactions between individuals' adjustments studied here.]

I close by mentioning some important related work. The results of the present paper complement those of Hines & Bishop (1983), who introduced learning effects into the inherited-strategies dynamics by allowing small individual gradient adjustments in parents' strategies immediately before their transmission to offspring. They treated the differences between the expected strategy frequencies faced by different individuals in a large population as negligible, aggregating individuals' adjustment processes and thereby eliminating the effects of the interactions between them studied here. They showed that the kind of learning effects they studied can increase strategy diversity over time, and can even lead to differences between the long-run population strategy frequencies and evolutionarily stable frequencies in the analogous inherited-strategies model. By way of comparison, the analysis presented here shows that, when learning effects predominate over fitness effects, the interactions between individuals' strategy adjustments alone, in any finite population, however large, can increase strategy diversity.

Selten (1988) studied the effect of individuals' anticipations of each other's strategy adjustments in a multi-species large-population model that otherwise resembles the model of Crawford (1974, 1985), showing that such anticipations can sometimes make mixed-strategy Nash equilibria stable.

Finally, Fudenberg & Kreps (1988) developed a game-theoretic model of learning by experimentation, with individuals much more sophisticated and better informed about the structure of their environment than assumed here, and used it to evaluate the intuition that persistent experimentation eventually leads individuals to a (suitably refined) Nash equilibrium.

I am grateful to Dennis Smallwood and the referees for helpful advice, to Yong-Gwan Kim for research assistance, and to the National Science Foundation for research support under grant SES 8703337.

NOTES

1. In what follows, a "large" population is one in which all individuals are assumed to face the same population strategy frequencies, and a "finite" population is one in which this is not assumed. "Mixed strategy" refers in general to randomized strategies, but does not exclude "pure" (that is, unrandomized) strategies.
2. Taylor & Jonker (1978) also considered discrete-time versions of the dynamics, for which the arguments in support of Maynard Smith's definition are weaker; see also Hines (1987: 241, 245–246).
3. A locally asymptotically stable equilibrium is one that has a neighborhood such that any trajectory that originates in the neighborhood converges to the equilibrium.
4. An equilibrium in the game played by matched pairs will be called a "Nash equilibrium" whenever this is necessary to preserve the distinction between game-theoretic equilibria and the equilibria of the population dynamics.
5. Equations (1)–(6) are inequalities.
6. Even if, by chance, $z^1 = \bar{z}$ for all i initially, random deviations of realized payoffs from their expected values in finite populations would quickly throw the system off, making aggregation impossible.
7. Equation (14) is an inequality.
8. The instability of mixed-strategy equilibria is described as only generic because the restrictions on payoff parameters under which it was proven for the Hawk–Dove game rule out non-generic parameter configurations for which $\kappa = 0$. The zero eigenvalues that result in such cases make the locally linearized dynamics unstable, but useless in determining whether the true non-linear dynamics are stable. (The parameter configurations that cause this problem are harder to identify for more general games, but still non-generic.) Working with a large population from the start of the analysis would evidently also yield zero eigenvalues, again making the local stability analysis inconclusive.
9. There are many surface differences between evolutionary games and the environments studied by Crawford (1974, 1985); this paper follows the conventions of

evolutionary game theory as closely as possible. The models of Crawford (1974, 1985) assume (by contrast) a fixed, finite population; possibly non-linear, absolute (instead of proportional) adjustment of mixed-strategy probabilities; discrete time; and repeated interaction of a single group of possibly asymmetric players, who ignore the effects of their current strategy adjustments on each other's future strategies (instead of anonymous, random pairing of identical players from a large population). The instability result survives all of these changes. It does depend on the form of the assumed learning rules, in that mixed-strategy adjustments must be differentiable functions of the payoff differences between pure strategies. Crawford (1985) gives references to the game-theoretic literature on learning.

10. Although the instability result establishing here depends on recognizing the finiteness of the population, it differs from other finite population-based criticisms of Maynard Smith's original ESS definition in the literature (see Schaffer, 1988; Maynard Smith, 1988; and the references given there) and, unlike those criticisms, remains significant in populations of indefinite size.

REFERENCES

Axelrod, R. 1984. *The Evolution of Cooperation.* New York: Basic Books.

Crawford, V. P. 1974. Learning the Optimal Strategy in a Zero-Sum Game. *Econometrica* **42**, 885.

Crawford, V. P. 1985. Learning Behavior and Mixed-Strategy Nash Equilibria. *J. econ. Behav. Organization* **6**, 69.

Cressman, R., and W. G. S. Hines. 1984. Correction to the Appendix of "Three Characterizations of Population Strategy Stability." *J. appl. Probab.* **21**, 213.

Eshel, I., and E. Akin. 1983. Coevolutionary Instability of Mixed Nash Solutions. *J. math. Biol.* **18**, 123.

Fudenberg, D., and D. Kreps. 1988. *A Theory of Learning, Experimentation, and Equilibrium in Games.* MIT and Stanford.

Harley, C. B. 1981. Learning the Evolutionarily Stable Strategy. *J. theor. Biol.* **89**, 611.

Hines, W. G. S. 1980a. Three Characterizations of Population Strategy Stability. *J. appl. Probab.* **17**, 333.

Hines, W. G. S. 1980b. Strategy Stability in Complex Populations. *J. appl. Probab.* **17**, 600.

Hines, W. G. S. 1981. Multi-Species Populations Models and Evolutionarily Stable Strategies. *J. appl. Probab.* **18**, 507.

Hines, W. G. S. 1987. Evolutionarily Stable Strategies: A Review of Basic Theory. *Theor. Pop. Biol.* **31**, 195 (erratum **33**, 114).

Hines, W. G. S., and D. T. Bishop. 1983. On Learning and the Evolutionarily Stable Strategy. *J. appl. Probab*, **20**, 689.

Houston, A. I., and B. H. Sumida. 1987. Learning Rules, Matching, and Frequency Dependence. *J. theor. Biol.* **126**, 289.

Maynard Smith, J. 1974. The Theory of Games and the Evolution of Animal Conflict. *J. theor. Biol.* **47**, 209.

Maynard Smith, J. 1982. *Evolution and the Theory of Games.* Cambridge, U.K.: Cambridge University Press.

Maynard Smith, J. 1988. Can a Mixed Strategy Be Stable in a Finite Population? *J. theor. Biol.* **130**, 247.

Maynard Smith, J., and G. A. Parker. 1976. The Logic of Asymmetric Contests. *Anim. Bchav.* **24**, 159.

Maynard Smith, J., and G. R. Price. 1973. The Logic of Animal Conflict. *Nature. Lond.* **246**, 15.

Schaffer, M. E. 1988. Evolutionarily Stable Strategies for a Finite Population and a Variable Contest Size. *J. theor. Biol.* **132**, 469.

Selten, R. 1988. *Anticipatory Learning in Two-Person Games. Discussion Paper* 4. Bonn: Institute fur Gesellschafts- und Wirtschaftswissenschaften, University of Bonn.

Sugden, R. 1986. *The Economics of Rights, Cooperation, and Welfare.* Oxford; New York: Basil Blackwell.

Taylor, P. and Jonker, L. 1978. Evolutionarily Stable Strategies and Game Dynamics. *Math. Biosci.* **40**, 145.

Zeeman, E. C. 1979. Population Dynamics from Game Theory, In: *Global Theory of Dynamical Systems Lecture Notes in Mathematics,* Vol. 819. Berlin; New York: Springer-Verlag.

Zeeman, E. C. 1981. Dynamics of the Evolution of Animal Conflicts. *J. theor. Biol.* **89**, 249.

8

Bayesian learning in games: A non-Bayesian perspective

J. S. JORDAN

Abstract

This paper is an exposition of some recent results in the Bayesian theory of learning in games. In a Bayesian learning process, players are assumed to have prior beliefs that are successively revised as play is repeated over time. This paper argues that for some such learning processes, the Bayesian methodology can be regarded as merely a heuristic device for constructing expectations that converge to the set of Nash equilibria for every game and, moreover, converge exponentially for most games. For this reason, Bayesian learning should be of interest even to those theorists who fear that it assumes more knowledge or rationality than is appropriate for a theory of learning.

1. INTRODUCTION

Economic equilibrium concepts typically stipulate that each economic agent chooses behavior that is individually maximizing given the equilibrium behavior of other agents. This raises the question of how each agent comes to know the equilibrium behavior of the others. If the individual characteristics of all agents are public knowledge, then presumably each agent could compute the equilibrium, provided that some selection criterion is adopted in the case of multiple equilibria. In most situations of economic interest, however, individual characteristics are private information not directly observable by others. This problem has led numerous authors to investigate the possibility that agents might learn from experience.

Suppose that the situation of interest is repeated over time while individual characteristics remain stationary. If the actions taken by each agent at each repetition are publicly observable, then agents might, through inductive inference or some process of trial and error, learn to form the correct expectations. If all of the agents except one are machines that take the same, possibly randomized

I would like to acknowledge especially helpful conversations with Larry Blume, David Easley, Ehud Kalai, and Yaw Nyarko, and the support of the National Science Foundation.

action each time, the inference problem for the single decision maker is straightforward. In the more typical case, however, learning influences the actions of each agent, and so influences what each agent subsequently learns. Thus economic models of learning typically generate complex interactive dynamics.

An equilibrium among individually maximizing agents is, in its most abstract form, the Nash equilibrium of a game. For this reason, much of the recent research in this area has focused on the question of whether the repeated play of a stationary game can lead the players to Nash equilibrium. Moreover, since the canonical model of expectation formation, at least in economic theory, is Bayes's theorem, the possibility of convergence to Nash equilibrium through Bayesian learning has attracted considerable attention (a partial survey is given by Blume and Easely 1992).

In Bayesian learning models, a player is assumed to begin with probabilistic beliefs about the sequences of actions the game will generate, and to revise those beliefs at each iteration in response to the actions observed to that point. The action chosen by a player at a particular iteration is optimal against the player's updated beliefs about the actions to be chosen by the other players. A number of recent results show that, if the initial beliefs satisfy certain conditions, then repeated play will lead to Nash equilibrium. This line of research is still at an early stage, but the results obtained thus far have been criticized for the restrictiveness of the required conditions on initial beliefs.

Under the usual subjectivist interpretation of Bayesian decisionmaking, initial beliefs are individual characteristics of the decision maker, much like the preferences and endowments of traders in a model of exchange equilibrium. From this point of view, learning results that depend on initial beliefs being the same for all players (Jordan 1991a, 1991b, 1992a) or being consistent, in a certain sense, with the actual distribution of action sequences (Kalai and Lehrer 1991a) have a narrow domain. On the positive side, this concern has led Nyarko (1991) to considerably expand the range of permissible beliefs, and on the negative side, it has led Blume and Easely (1992) to conclude that Bayesian learning "cannot provide a foundation for Nash equilibrium."

The object of the present paper is to show that some of the results in Bayesian learning have an alternative non-Bayesian interpretation in which the initial beliefs are only incidentally involved. To paraphrase Diaconis and Freedman (1986, p. 66), we will view the formation of expectations via the Bayesian revision of prior beliefs as a "powerful heuristic engine" for the construction of learning processes. Of course, this interpretation is tenable only if the performance of a learning process can be evaluated independently of the prior beliefs from which it was derived. In particular, expectations must be shown to converge to Nash equilibrium for all games, or at least an "objectively" large set of games, rather than merely a set of games that is likely according to the initial beliefs. Second, even if initial beliefs have no asymptotic influence, they will generally influence the path to equilibrium, so it is desirable that the rate of convergence be rapid. It will be shown in Section 2 that a certain class of Bayesian learning procedures meets both criteria.

Formal definitions and results will be stated in Section 2, but a small amount of notation is necessary to make the present discussion more precise. Suppose that the actions available to player p at each iteration lie in a finite set S_p of "pure strategies." There are n players, and, following iteration t, each player observes the n-tuple $s_t = (s_{1t}, \ldots, s_{nt})$ of chosen actions. A learning process for player p is a sequence of functions e_{pt}, which associate with any observed finite history of play $h_t = (s_1, \ldots, s_t)$ an expectation $e_{pt}(h_t) \in \Delta(S_{-p})$, where $\Delta(S_{-p})$ is the set of probability distribution over $(n-1)$-tuples $s_{-p} = (s_1, \ldots, s_{p-1}, s_{p+1}, \ldots, s_n)$ of actions to be chosen by the other players in period $t+1$. After each iteration t, each player p receives the payoff $\pi_p(s_t)$. Player p knows the payoff function π_p, which remains fixed through time, but does not know the payoff function of any other player. Player p chooses s_{pt} in period t to maximize the expected value of π_p given the expectations $e_{p(t-1)}(h_{t-1})$ about the other players' actions. The payoff functions $(\pi_p)_{p=1}^n$ determine the set of Nash equilibria of the one-shot game, so the learning question is whether this sort of repetition of the one-shot game causes the expectations sequence $\{e_{pt}(h_t)\}_{t=1}^\infty$ to approach the set of Nash equilibrium expectations.

The possibility of convergence to Nash equilibria of the one-shot game depends on the assumption that players choose actions at each iteration to maximize current expected payoffs. The case in which players seek to maximize the expected discounted sum of payoffs over future iterations was first investigated by Kalai and Lehrer (1991a, 1991b). They proved that if the players' initial beliefs are probability distributions on infinite paths of opponents' future actions such that, given the true payoff functions $(\pi_p)_p$ (and discount factors), the actual distribution of best-response histories is absolutely continuous with respect to each player's initial belief, then the players' behavior strategies will approach Nash equilibrium play. More precisely, they showed that behavior strategies approach the set of "subjective equilibria," and that behavior strategies that are nearly in subjective equilibrium also approximate the play of a repeated-game Nash equilibrium. Jordan (1991b) subsequently proved that players' expectations approach the set of Nash equilibria with subjective probability one, under the assumptions that players initially believe that the payoff functions π_p (and discount factors) are generated independently, according to distributions μ_p, which are believed in common by all players, and that expectations are determined as a Bayesian Nash equilibrium in behavior strategies. Nyarko (1991) proved that the common-beliefs assumption could be replaced by a much more general belief hierarchy satisfying a certain mutual consistency condition. However, none of the results obtained to date for the case in which players maximize the expected discounted sum of future payoffs seems amenable to a non-Bayesian interpretation.

The problem of convergence to one-shot Nash equilibria, from a non-Bayesian perspective, is one of finding expectation functions e_{pt} that perform well for a large class of games. This problem is solved by the following Bayesian construction. Suppose that all players believe that the game $(\pi_p)_{p=1}^p$ is generated by drawing each π_p independently from a distribution μ_p of payoff functions for player p. Viewing each π_p as the "type" of player p, the players' expectations about each other's first period actions can be derived as a Bayesian Nash equilibrium. After the first period actions $s_1 = (s_{11}, \ldots, s_{n1})$ are observed, each player knows that player p chose s_{p1} to maximize the expected value of the "true" π_p. This knowledge enables each μ_p to be revised, leading to the determination of the expectations $e_{p1}(s_1) \in \Delta(S_{-p})$ as a Bayesian Nash equilibrium for the revised beliefs, and so on. This learning process is termed "sophisticated Bayesian learning" because of the intricate deductive inferences on which it is based. Theorem 2 of Section 2 establishes that if the prior beliefs are sufficiently uniform, then expectations are asymptotically Nash for *every* game $(\pi_p)_p$.

There are two important qualifications to Theorem 2. First, unless the game has a unique Nash equilibrium, the expectations sequence $\{e_{pt}(h_t)\}_{t=1}^{\infty}$ is not asserted to have a unique limit. Rather, Theorem 2 asserts that every cluster point of the expectations sequence is a Nash equilibrium. Second, although expectations are asymptotically Nash, the actions played over time need not be asymptotically Nash. If expectations converge to a strict pure-strategy Nash equilibrium, then, of course, the pure strategies $(s_{1t}, \ldots, s_{nt})$ played over time must converge to the same equilibrium. However, no such implication holds if expectations converge to a mixed-strategy Nash equilibrium. Indeed, Theorem 1 states that for any expectation functions e_{pt}, except for a set of games having Lebesgue measure zero, every player's maximizing action in each period is a unique pure strategy. Thus, there is no hope for a learning process that generally leads expected payoff maximizing players to asymptotically *play* mixed strategies. On the other hand, Theorems 3 and 3' establish that except for a set of games with Lebesgue measure zero, sophisticated Bayesian learning asymptotically purifies mixed-strategy equilibria in the sense that the pure strategies played converge in frequency to the same Nash limits as the expectations sequence.

Theorem 4 states that for "regular" games, the expectations generated by sophisticated Bayesian learning approach Nash equilibrium exponentially. Since the set of nonregular games is again a set of Lebesgue measure zero, Theorem 4 provides the desired speed of convergence result.

Section 3 is a detailed description of a sophisticated Bayesian learning process for 2×2 games. For 2×2 games, the construction of expectations is geometrically transparent.

As befits an exposition, there is nothing in this paper that is really new. Theorems 2 and 4 have been proved elsewhere, and the proofs are not repeated here. Proofs are given for Theorem 1, which is a new version of a previous result (Jordan 1992a, Theorem 2.2), and Theorems 3 and 3', which are essentially anticipated by Nyarko (1992, Theorem 9.6).

2. GENERAL DEFINITIONS AND RESULTS

Definition 2.1 *There are n players, $n \geq 2$, indexed by the subscript p. For each $1 \leq p \leq n$, let S_p be a finite set, and let $S = \Pi_{p=1}^n S_p$. For each p, let $S_{-p} = \Pi_{q \neq p} S_q$, with generic element $s_{-p} = (s_1, \ldots, s_{p-1}, s_{p+1}, \ldots, s_n)$. Each player p is characterized by a payoff function $\pi_p : S \to R$. The space of possible payoff functions for each player is the unit sphere in R^s, that is, $B = \{\pi \in R^s : (\Sigma_{s \in S} \pi(s)^2)^{1/2} = 1\}$. Given the strategy sets S_p, a normal form game is completely specified by an n-tuple of payoff functions $(\pi_p)_p \in B^n$, so let $G = B^n$ denote the space of all games, with generic element $(\pi_p)_p$.*

A probability distribution σ on S is a product distribution if $\sigma = \sigma_1 \times \cdots \times \sigma_n$, where each σ_p is the marginal distribution on S_p. That is, $\sigma(s) = \sigma_1(s_1) \cdots \sigma_n(s_n)$ for each $s \in S$. A Nash equilibrium for a game $(\pi_p)_p \in G$ is a product distribution σ^* on S such that for each p, σ_p^* maximizes $\Sigma_{s_p} \sigma_p(s_p) \Sigma_{s_{-p}} \pi_p(s_p, s_{-p}) \sigma_{-p}^*(s_{-p})$ over the set of probability measures σ_p on S_p, where $\pi_p(s_p, s_{-p}) = \pi_p(s_1, \ldots, s_{p-1}, s_p, s_{p+1}, \ldots, s_n)$, and σ_{-p}^* is the distribution $\Pi_{q \neq p} \sigma_q^*$ on S_{-p}. For each $(\pi_p)_p \in G$, let $N((\pi_p)_p)$ denote the set of Nash equilibria.

The normalization of payoff functions to B will be convenient below, because B has finite Lebesgue measure. In Jordan (1991a), B was taken to be the unit ball in R^s for the additional convenience of convexity. A further normalization will be used for 2×2 games in Section 3. Of course, none of these normalizations are essential to the analysis or results.

Throughout this paper, players will be assumed to choose strategies as best responses to their expectations of the strategies of the other players. Under this assumption, a learning mechanism for each player p is a sequence of expectations functions, e_{pt}, mapping the history of play observed through period t to a forecast of the strategies that the other players will play in period $t + 1$.

Definition 2.2 *For each $t \geq 1$, let $H_t = \Pi_{\tau=1}^t S$, with generic element $h_t = (s_1, \ldots, s_t)$, and let H_0 denote the one-element set $\{*\}$. Let $H_\infty = \Pi_{\tau=1}^\infty S$, with generic element $h_\infty = (s_1, \ldots)$. Given any $t \geq 0$ and any $h_\infty = (s_1, \ldots)$, let*

153

$h_{\infty|t} \in H_t$ *denote the t-period truncation of h_∞, that is, $h_{\infty|t} = (s_1, \ldots, s_t)$ if* $t \geq 1$ *and* $h_{\infty|0} = *$. *A subhistory of an infinite history h_∞ is a sequence of truncations $\{h_{\infty|t_k}\}_{k=1}^\infty$ such that $t_{k+1} > t_k$ for all k.*

A learning process for player p is a sequence $e_p = \{e_{pt}\}_{t=0}^\infty$ of functions $e_{pt} : H_t \to \Delta(S_{-p})$, *where $\Delta(S_{-p})$ is the set of probability distributions on* S_{-p}. *A learning process is an n-tuple $(e_p)_{p=1}^n$ of learning processes for the respective players.*

In specifying H_t as the domain of e_{pt}, we assume that player p remembers the entire history of play, $(s_1, \ldots, s_t)$, through period t, but has no direct knowledge of the payoff function π_q of any other player q. In excluding π_p from the domain of e_{pt}, we assume that player p's knowledge of π_p does not reveal any information about π_q for $q \neq p$.

A sequence of expectation functions $\{e_{pt}\}_{t=0}^\infty$ can be derived from a subjective probability distribution ψ_p on H_∞ as a sequence of conditional probabilities:

$$e_{pt}(h_t)(s_{-p}) = \psi_p(s_{-p(t+1)} = s_{-p}|h_{\infty|t} = h_t),$$

where the conditional probability is arbitrary given finite histories outside the support of ψ_p. More precisely, $e_{pt}(h_t)$ is an arbitrary element of $\Delta(S_{-p})$ if $\psi_p(\{h_\infty : h_{\infty|t} = h_t\}) = 0$. Conversely, given any sequence of expectation functions $\{e_{pt}\}_{t=0}^\infty$, one can construct a distribution ψ_p on H_∞ yielding the given sequence as conditional probabilities. For example, for each $s \in S$, let $\psi_p(s_1 = s) = (1/\#S_p)e_{p0}(s_{-p})$, and for each $t \geq 1$, each h_t, and each s, let

$$\psi_p(s_{t+1} = s|h_{\infty|t} = h_t) = (1/\#S_p)e_{pt}(h_t)(s_{-p}).$$

That is, ψ_p is constructed iteratively by using each e_{pt} as the conditional distribution generating $s_{-p(t+1)}$ and specifying that s_{pt} is drawn independently each period from the uniform distribution on S_p.

Thus any learning process $(e_p)_p$ can be interpreted as "Bayesian learning" based on subjective probability distributions $(\psi_p)_p$ on H_∞. The Bayesian interpretation of learning appears to place no restrictions on learning dynamics except the restriction that each s_{pt} must be chosen as a best response to player p's expectations, given the payoff function π_p. If expectations were completely arbitrary, this restriction would only prevent the play of strictly dominated strategies. However, since the expectation functions $e_{pt}(\cdot)$ do not have π_p as an argument, the "best-response" assumption constrains the pattern of strategy choices s_{pt} across payoff functions π_p.

Definition 2.3 *A best-response history (BRH) for a game $(\pi_p)_p$ and a learning process $(e_P)_p$ is an infinite history $h_\infty = (s_1, \ldots)$ satisfying, for each $t \geq 0$ and each p, $s_{p(t+1)}$ maximizes $\Sigma_{s_{-p}}\pi_p(\cdot, s_{-p})e_{pt}(h_{\infty|t})(s_{-p})$.*

A best-response history can be constructed period by period in the obvious way, so it is clear that a BRH exists for every game and learning process. It is less clear, but straightforward to prove, that for any given learning process, the BRH is unique for almost every game.

Theorem 1 *Given a learning process $(e_p)_p$, there is a set of games $G^0((e_p)_p)$ with Lebesgue measure zero such that if $(\pi_p)_p \notin G^0((e_p)_p)$ there is a unique BRH for $(e_p)_p$ and $(\pi_p)_p$.*

PROOF: For each p and each $\sigma_{-p} \in \Delta(S_{-p})$, let $C_p(\sigma_{-p}) = \{\pi_p \in B$: for some $s_p, s_p' \in S_p$ with $s_p \neq s_p'$, $\Sigma_{s_{-p}}(\pi(s_p, s_{-p}) - \pi(s_p', s_{-p}))\sigma_{-p}(s_{-p}) = 0\}$. Then $C_p(\sigma_{-p})$ has Lebesgue measure zero in B. For each $t \geq 0$, let $K_{pt} = \cup\{C_p(\sigma_{-p}) : \sigma_{-p} \in e_{pt}(H_t)\}$. Since H_t is a finite set for each t, K_{pt} has Lebesgue measure zero as a finite union of sets of measure zero. Let $K_p = \cup_{t \geq 0} K_{pt}$, and let $G^0 = \{(\pi_p)_p \in B^n : \pi_p \in K_p$ for some $p\}$. Then G^0 has Lebesgue measure zero in B^n. Moreover, if $(\pi_p)_p \notin G^0$, then for each p, t, and each $h_t \in H_t$, the function $\Sigma_{s_{-p}} \pi_p(\cdot, s_{-p}) e_{pt}(h_t)(s_{-p})$ is $1 - 1$ on S_p, and thus has a unique maximum. $\qquad\qquad\square$

Theorem 1 implies that even if the expectations generated by a learning process $(e_p)_p$ are nondegenerate, the best response history is generically a unique sequence of pure strategies. That is, for almost all games, the "true" probability distribution on H_∞ is degenerate. Since the set of games having a mixed-strategy Nash equilibrium as the unique Nash equilibrium has positive Lebesgue measure (it has a nonempty interior), Theorem 1 immediately implies the nonexistence of any learning process for which best-response strategies converge to $\mathcal{N}((\pi_p)_p)$ for all games $(\pi_p)_p$. That is, players cannot generally "learn to play" mixed-strategy Nash equilibria. Of course Theorem 1 does not preclude the general convergence of expectations or of the empirical frequencies of best-response strategies to Nash equilibrium.

Definition 2.4 *Given a learning process $(e_p)_p$ and an infinite history h_∞, we say that expectations are asymptotically consistent if for each $t \geq 0$, there is some $\sigma_t \in \Delta(S)$ such that for every player p,*

$$\lim_{t \to \infty} ||e_{pt}(h_{\infty|t}) - \sigma_{-pt}|| = 0,$$

where σ_{-pt} is the marginal distribution of σ_t on S_{-p}. If expectations are asymptotically consistent, let $L((e_p)_p, h_\infty)$ denote the set of all cluster points of the sequence $\{\sigma_t\}_{t=0}^\infty$. Note that since $\Delta(S)$ is compact, $L((e_p)_p, h_\infty) \neq \emptyset$. A learning process $(e_p)_p$ is generally convergent if for every game $(\pi_p)_p$ there is a best-response history h_∞ for $(\pi_p)_p$ and $(e_p)_p$ such that expectations are asymptotically consistent and $L((e_p)_p, h_\infty) \subset \mathcal{N}((\pi_p)_p)$.

The asymptotic consistency of expectations requires that the expectations of the different players be asymptotically compatible with a single distribution σ_t in each period t (since $\Delta(S)$ can be represented as the unit simplex in $R^{\#s}$, the norm $\| \cdot \|$ can be taken to be the Euclidean norm on $R^{\#s}$). Asymptotic consistency is automatically satisfied, with $\sigma_{-pt} = e_{pt}(h_{\infty|t})$ for all t, if there are only two players, or if the expectation functions $\{e_{pt}\}_{t=0}^{\infty}$ for all players are derived as conditional expectations from a common prior distribution ψ on H_{∞} (given consistent definitions of the expectation functions for histories not in the support of ψ).

General convergence does not always require the sequence $\{\sigma_t\}$ to converge. If a game $(\pi_p)_p$ has multiple Nash equilibria, then the sequence $\{\sigma_t\}$ may have several or all of these equilibria as cluster points. However, there can be no cluster points that are not Nash equilibria. In other words, the sequence $\{\sigma_t\}$ need not have a single limit but must converge to a subset of $\mathcal{N}((\pi_p)_p)$ for some best-response history h_{∞}. If a game has multiple best-response histories, some histories may yield nonequilibrium cluster points. However, Theorem 1 indicates that best-response histories are typically unique.

Definition 2.5 *For each p, each $t \geq 1$, and each $h_t = (s_1, \ldots, s_t)$, define the empirical frequency $f_{pt}(h_t)(s_p) = \#\{\tau \leq t : s_{p\tau} = s_p\}/t$ for each $s_p \in S_p$. Any learning process $(e_p)_p$ such that for each p and each $t \geq 1$,*

$$e_{pt}(h_t)(s_{-p}) = \Pi_{q \neq p} f_{qt}(h_t)(s_q)$$

for all $h_t \in H_t$ and all $s_{-p} \in S_{-p}$ will be called a fictitious-play learning process. The initial expectations e_{p0} for each p are arbitrary.

A fictitious-play learning process $(e_p)_p$ is automatically asymptotically consistent for every history h_{∞} (let $\sigma_t = \Pi_{p=1}^{n} f_{pt}(h_{\infty|t})(\cdot)$ for each t). However, the well-known example of Shapley (1964) (see also Jordan 1992b) shows that fictitious play is not generally convergent.

To date, the only known generally convergent learning processes are the "sophisticated Bayesian" learning processes defined below.

Definition 2.6 *For any p, a Borel probability distribution μ_p is uniformly absolutely continuous (u.a.c.) if μ_p has a density function $g_p : B \to R$ with respect to Lebesgue measure such that for some constant $c_p > 1$,*

$$c_p > g_p(\pi_p) \geq 1/c_p \quad for\, every \quad \pi_p \in B.$$

Let μ be a Borel probability distribution on G such that $\mu = \mu_1 \times \cdots \times \mu_n$ where each μ_p is u.a.c. Let φ be a Borel probability distribution on $G \times H_{\infty}$ satisfying

(i) the marginal distribution on G agrees with μ.

Let ψ denote the marginal distribution of φ on H_∞ and for each $t \geq 0$, define the conditional probabilities $e_t : H_t \to \Delta(S)$ by

$$e_t(h_t)(s) = \psi(s_{t+1} = s | h_{\infty|t} = h_t),$$

and for each p, define $e_{pt} : H_t \to \Delta(S_{-p})$ by setting $e_{pt}(h_t)$ equal to the marginal distribution on S_{-p} induced by $e_t(h_t)$. If, in addition to (i), φ satisfies

(ii) for every $((\pi_p)_p, h_\infty) \in$ supp φ, h_∞ is a best-response history for $(e_p)_p$, and $(\pi_p)_p$,

where supp φ denotes the support of φ, then $(e_p)_p$ is a sophisticated Bayesian learning process.

Condition (ii) is a "rational expectations equilibrium" condition, stating that expectations are consistent with expected payoff maximization. In Jordan (1991b, 2.4), the distribution φ was termed a *Bayesian strategy process* (BSP). Given any product distribution $\mu = \mu_1 \times \cdots \times \mu_n$, a BSP could be constructed period by period from Bayesian Nash equilibria for successively updated prior distributions on G (Jordan 1991b, 2.5 and Proposition 2.6). Hence many sophisticated Bayesian learning processes exist. The uniform-absolute-continuity assumption is needed to ensure general convergence (Jordan 1991b, Example 3.11). The following result is a direct implication of Jordan (1991b, Corollary 3.10).

Theorem 2 *Every sophisticated Bayesian learning process is generally convergent.*

The main result of Kalai and Lehrer (1991a, Theorem 2), applied to the present case of myopic payoff functions, implies a stronger form of convergence on a narrower domain. Given a learning process $(e_p)_p$, let $(\psi_p)_p$ be distributions on H_∞ for which the respective expectations functions are conditional expectations. For example, the ψ_p's could be constructed as described following Definition 2.2 above. Given a game $(\pi_p)_p$, let ψ be a distribution on H_∞ such that every $h_\infty \in$ supp ψ is a best-response history for $(\pi_p)_p$ and $(e_p)_p$. Suppose that ψ is absolutely continuous with respect to each ψ_p. Then the main result of Kalai and Lehrer (henceforth K–L) states that the best response strategies, as well as the expectations of each player, converge to a subset of $\mathcal{N}((\pi_p)_p)$.

Theorem 1, above, indicates both the strength of the K–L conclusion and the restrictiveness of their hypothesis. Since ψ is typically degenerate, the K–L absolute-continuity condition will typically require that each ψ_p contain what K–L call a "grain of truth," that is, an atom of probability at the unique best-response history. There is an interesting class of learning processes and games that satisfy this condition. Suppose that, in the above definition of sophisticated Bayesian learning, the assumption that each μ_p is u.a.c. is replaced by the assumption that each μ_p has countable support. Then K–L (Theorem 1.1 and Theorem 2.1) show that their absolute-continuity condition will be satisfied by every game $(\pi_p)_p$ in the (countable) support of μ. In other words, if the players' knowledge of the true game is increased to this extent, then they can "learn to play" even mixed-strategy Nash equilibria.

Theorem 2 says nothing about the asymptotic behavior of the best-response history h_∞. Of course, if expectations converge to a strict pure-strategy Nash equilibrium, then the best-response strategies s_t must converge to the same equilibrium. However, if expectations converge to a mixed-strategy Nash equilibrium, then nothing can be concluded from Theorem 2 about the best-response strategies. Fortunately, an argument based solely on the martingale-convergence theorem, with no essential reference to our game-theoretic context, shows that the empirical frequency of any strategy n-tuple almost always approaches its average expected probability. Thus, if expectations converge to a mixed-strategy equilibrium, the best-response strategies converge to the same equilibrium in frequency, even though the players typically will never actually "play" mixed strategies.

Definition 2.7 *For each* $t \geq 1$, *define the empirical frequency distribution* $f_t : H_t \to \Delta(S)$ *by*

$$f_t(h_t)(s) = (1/t)\#\{\tau \leq t : s_\tau = s\}$$

for each $s \in \dot{S}$.

Theorem 3 *Let* $(e_p)_p$ *be a sophisticated Bayesian learning process, and let* $\{e_t\}_{t=0}^\infty$ *be the sequence of functions specified in Definition 2.6 (from which the functions* e_{pt} *are derived as marginal probabilities on* S_{-p}*). Then there is a set* $G'((e_p)_p) \subset G$ *with Lebesgue measure zero such that for each* $(\pi_p)_p \notin G'((e_p)_p)$, *if* h_∞ *is the unique best-response history for* $(\pi_p)_p$ *and* $(e_p)_p$, *then*

$$(*) \qquad \lim_{t \to \infty} \| f_t(h_{\infty|t}) - (1/t) \sum_{\tau=1}^{t} e_{\tau-1}(h_{\infty|\tau-1}) \| = 0.$$

Corollary 1 *Under the hypothesis and notation of Theorem 3,*

$$\liminf_{t \to \infty} \{ \| f_t(h_{\infty|t}) - \sigma \| : \sigma \in co\mathcal{N}((\pi_p)_p) \} = 0,$$

where co denotes the convex hull. In particular, if the game $(\pi_p)_p$ *has a unique Nash equilibrium* σ, *then* $f_t(h_{\infty|t}) \to \sigma$.

PROOF: Let μ, φ, and ψ be associated with $(e_p)_p$ as in Definition 2.6. Given $s \in S$, define the indicator functions $\chi_t : H_\infty \to \{0, 1\}$ by

$$\chi_t(h_\infty) = \begin{cases} 1 & \text{if } s_t = s; \text{ and} \\ 0 & \text{otherwise,} \end{cases}$$

for each $t \geq 1$; and define the random variables $m_t : H_\infty \to R$ by

$$m_t(h_\infty) = \sum_{\tau=1}^{t} (1/\tau)[\chi_\tau(h_\infty) - E\{\chi_\tau(h_\infty)|h_{\infty|\tau-1}\}]$$

$$= \sum_{\tau=1}^{t} (1/\tau)[\chi_\tau(h_\infty) - e_{\tau-1}(h_{\infty|\tau-1})(s)],$$

where $E\{\cdot|\cdot\}$ is the conditional-expectations operator induced by ψ. Then the sequence $\{m_t(\cdot)\}_{t=1}^{\infty}$ is a martingale. In order to apply the martingale-convergence theorem (e.g., Breiman 1968, Theorem 5.14, p. 89), it suffices to show that

$$\limsup_{t \to \infty} E m_t^2(\cdot) < \infty.$$

Note that if $\tau < \tau'$, then by iterating conditional expectations,

$$E\{(1/\tau)[\chi_\tau(h_\infty) - E\{\chi_\tau(h_\infty)|h_{\infty|\tau-1}\}](1/\tau')[\chi_{\tau'}(h_\infty)$$

$$- E\{\chi_{\tau'}(h_\infty)|h_{\infty|\tau'-1}\}]\} = E\{[\chi_\tau(h_\infty) - E\{\chi_\tau(h_\infty)|h_{\infty|\tau-1}\}]$$

$$E\{[\chi_{\tau'}(h_\infty) - E\{\chi_{\tau'}(h_\infty)|h_{\infty|\tau'-1}\}|h_{\infty|\tau}]\}/(\tau\tau')\} = 0.$$

Therefore,

$$E m_t^2(\cdot) = \sum_{\tau=1}^{t} (1/\tau^2) E\{\chi_\tau(h_\infty) - E\{[\chi_\tau(h_\infty)|h_{\infty|\tau-1}]\}\}^2$$

$$\leq \sum_{\tau=1}^{t} (1/\tau^2)(1/4) < (1/4) \sum_{\tau=1}^{\infty} (1/\tau^2) < \infty.$$

It follows from the martingale-convergence theorem that there is a random variable $m^* : H_\infty \to R$ and a subset $H_\infty' \subset H_\infty$ with $\psi(H_\infty') = 1$ such that $m_t(h_\infty) \to m^*(h_\infty)$ for every $h_\infty \in H_\infty'$. For any $h_\infty \in H_\infty'$, since the series $\Sigma_{\tau=1}^{\infty}(1/\tau)(\chi_\tau(h_\infty) - e_{\tau-1}(h_{\infty|\tau-1})(s))$ converges, it follows that $\lim_{t \to \infty} 1/t \Sigma_{\tau=1}^{t}(\chi_\tau(h_\infty) - e_{\tau-1}(h_{\infty|\tau-1})(s)) = 0$. Hence $\lim_{t \to \infty} ||f_t(h_{\infty|t}) (s) - (1/t)\Sigma_{\tau=1}^{t} e_{\tau-1}(h_{\infty|\tau-1})(s)|| = 0$ for each $h_\infty \in H_\infty'$. Since S is a finite set, the above argument can be applied separately to each $s \in S$ to obtain s set H_∞'', as the intersection of the respective sets H_∞', with the properties that $\psi(H_\infty'') = 1$ and $(*)$ holds for every $h_\infty \in H_\infty''$.

Let μ be the probability distribution on G specified in the definition of a sophisticated Bayesian learning process, and let $G^0 = G^0((e_p)_p) \subset G$ be given by Theorem 1. Since each μ_p is u.a.c., $\mu(G^0) = 0$. Let $G'' = \{(\pi_p)_p \in G\backslash G^0$: for some $h_\infty \in H''_\infty, ((\pi_p)_p, h_\infty) \in \text{supp } \varphi\}$. Since μ is the marginal of φ on G and ψ is the marginal of φ on H_∞, $\psi(H''_\infty) = 1$ implies that $\mu(G'') = 1$. Since $G \cap G^0 = \emptyset$, Theorem 1 and 2.6 (ii) imply that for any $(\pi_p)_p \in G''$ and any h_∞ with $((\pi_p)_p, h_\infty) \in \text{supp } \varphi$, h_∞ is the unique BRH for $(\pi_p)_p$ and $(e_p)_p$. Finally, let $G'((e_p)_p) = G\backslash G''$. Since $\mu(G'') = 1$ and each μ_p is u.a.c., $G'((e_p)_p)$ has Lebesgue measure zero. $\qquad\square$

The corollary follows directly from $(*)$ and Theorem 2.

The purpose of stating the theorem in terms of the joint frequencies $f_t(h_{\infty|t})$ and expectations $e_t(h_{\infty|t})$ rather than the marginals on each S_{-p} is to obtain a stronger conclusion. Joint frequency convergence implies convergence of the marginals, but the converse implication is false. An example is described in Young (1991, p. 9) and Jordan (1992b, Section 3) of a two-player game in which fictitious-play learning leads to convergence of the marginal frequencies to Nash equilibrium, but the joint frequency distribution is not asymptotically Nash.

There are two principal limitations to the theorem and corollary. First, if a game $(\pi_p)_p$ has multiple Nash equilibria that are limits of distinct subhistories of h_∞, then the "average" expectations $(1/t)\Sigma_{\tau=1}^t e_{\tau-1}(h_{\infty|\tau-1})$ approach the convex hull of $\mathcal{N}((\pi_p)_p)$, but need not approach $\mathcal{N}((\pi_p)_p)$. Second, Theorem 3 violates the non-Bayesian prohibition against references to prior probabilities. A strategy frequency, like the frequency of zeros in the binary expansion of a random number, has the desired behavior "almost always" but not always. The fact that $(*)$ is violated only on a set of Lebesgue measure depends on the fact that since each μ_p is u.a.c., any set with μ-probability zero has Lebesgue measure zero. Nonetheless, it does not appear that any stronger convergence assertion for empirical strategy frequencies could be proved, even under the Kalai and Lehrer assumption that each μ_p has countable support.

While Theorem 3 and Corollary 1 appear to exhaust the empirical content of Theorem 2, there remains a conceptual gap in the case of multiple Nash equilibria. Suppose that the set of cluster points $L((e_p)_p, h_\infty)$ contains several Nash equilibria and that along a subhistory $\{h_{\infty|t_k}\}_{k=1}^\infty$, $e_t(h_{\infty|t_k}) \to \sigma \in L((e_p), h_\infty)$. One would like the frequency distribution of best-response strategies, restricted to the subsequence $\{s_{t_k+1}\}_{k=1}^\infty$, to converge to σ also. Of course, frequency convergence could not be obtained for every subhistory with expectations converging to σ. If σ is a mixed-strategy equilibrium, then from any subhistory for which the strategy frequencies converge to σ, one can select a further subhistory along which only a particular strategy $s \in \text{supp } \sigma$ is played.

Nonetheless, if a subhistory is, as defined below, maximal with respect to the property that expectations converge to σ, then Theorem 3' gives the desired frequency convergence. A result very close to Theorem 3' was obtained by Nyarko (1992, Theorem 9.6) using essentially the same martingale-convergence argument used here.

Definition 2.8 *Given a sequence $\{\sigma_t\}_{t=1}^{\infty} \subset \Delta(S)$ and a distribution $\sigma \in \Delta(S)$, a subsequence $\{\sigma_{t_k}\}_{k=1}^{\infty}$ is maximal for σ if*

(i) $\lim_{k \to \infty} \sigma_{t_k} = \sigma$; *and*
there is some $\varepsilon > 0$ such that
(ii) for every t, either $\|\sigma_t - \sigma\| > \varepsilon$ or $t = t_k$ for some k.
A sequence $\{s_k\}_{k=1}^{\infty} \subset S$ is said to converge in frequency to a distribution $\sigma \in \Delta(S)$ if

$$\lim_{k \to \infty} (1/k) \#\{k' \le k : s_{k'} = s\} = \sigma(s)$$

for each $s \in S$.

Theorem 3' *Under the hypothesis of Theorem 3, there is a set $G'((e_p)_p) \subset G$ with Lebesgue measure zero such that if $(\pi_p)_p \notin G'((e_p)_p)$ and h_∞ is the unique best-response history for $(\pi_p)_p$ and $(e_p)_p$, then for any $\sigma \in L((e_p)_p, h_\infty)$ and any subhistory $\{h_{\infty|t_k}\}_{k=1}^{\infty}$ such that $\{e_{t_k}(h_{\infty|t_k})\}_{k=1}^{\infty}$, as a subsequence of $\{e_t(h_{\infty|t})\}_{t=1}^{\infty}$, is maximal for σ, it follows that the sequence $\{s_{t_k+1}\}_{k=1}^{\infty}$ converges in frequency to σ. Moreover, for each player p,*

$$\lim_{k \to \infty} (1/k)\pi_p(s_{t_k+1}) = \sum_s \pi_p(s)\sigma(s).$$

PROOF: Let $\{b_r\}_{r=1}^{\infty}$ be a countable collection of open balls in $R^{\#S}$ such that $\{b_r \cap \Delta(S)\}_{r=1}^{\infty}$ is a basis for the Euclidean topology on $\Delta(S)$. For each r, t, define $\lambda_t^r : H_\infty \to [0, 1]$ by

$$\lambda_t^r(h_\infty) = \begin{cases} 1 & \text{if } e_t(h_{\infty|t}) \in b_r; \text{ and} \\ 0 & \text{otherwise,} \end{cases}$$

define $T_t^r : H_\infty \to R$ by $T_t^r(h_\infty) = \#\{\tau \le t : \lambda_\tau^r(h_\infty) = 1\}$, define $f_t^r : H_\infty \times S \to [0, 1]$ by $f_t^r(h_\infty, s) = \#\{\tau \le t : s_\tau = s \text{ and } \lambda_{\tau-1}^r(h_\infty) = 1\}/T_{t-1}^r(h_\infty)$, where we adopt the convention that $0/0 = 0$. Then the proof of Theorem 3 can be modified slightly to show that for each r, there is a set $G''((e_p)_p)$ such that if $(\pi_p)_p \notin G''((e_p)_p)$ and h_∞ is the unique BRH for $(\pi_p)_p$ and $(e_p)_p$, then for all $s \in S$,

$$(*_r) \lim_{t \to \infty} \left| f_t^r(h_\infty, s) - \left(\sum_{\tau=1}^{t} \{e_{\tau-1}(h_{\infty|\tau-1})(s) : \lambda_{\tau-1}^r(h_\infty) = 1\} \right) \middle/ \right.$$
$$\left. T_{t-1}^r(h_\infty) \right| = 0 \text{ if } T_{t-1}^r(h_\infty) \to \infty$$

161

for all $s \in S$. Let $G'((e_p)_p) = \cap_{r=1}^{\infty} G''((e_p)_p)$. Now let $(\pi_p)_p \in G'((e_p)_p)$, let h_∞ be the unique BRH for (π_p) and $(e_p)_p$, and let $\{e_{t_k}(h_{\infty|t_k})\}_{k=1}^{\infty}$ be maximal for some $\sigma \in L((e_p)_p, h_\infty)$. Then for some r with $\sigma \in b_r, e_{t_k}(h_{\infty|t_k}) \notin b_r$ for only finitely many k, and for every $t, e_t(h_{\infty|t}) \in b_r$ only if $t = t_k$ for some k. Since $e_{t_k}(h_{\infty|t_k}) \rightarrow \sigma$,

$$\sum_{\tau=1}^{t} \{e_{\tau-1}(h_{\infty|\tau-1})(\cdot) : \lambda_{\tau-1}(h_\infty) = 1\}/T_{t-1}^r(h_\infty) \rightarrow \sigma,$$

so $(*_r)$ implies that the sequence $\{s_{t_k+1}\}_{k=1}^{\infty}$ converges in frequency to σ. The final assertion of the theorem follows immediately. $\square$

Theorem 2 implies that any $\sigma \in L((e_p)_p, h_\infty)$ is a Nash equilibrium for $(\pi_p)_p$. Hence, Theorem 3' implies that along maximal subsequences for σ, the best-response strategies converge in frequency to Nash equilibrium and the players' average payoffs converge to the Nash equilibrium expected payoffs.

It should be noted that for some games, maximal subsequences are not guaranteed to exist. If $\mathcal{N}((\pi_p)_p)$ is a connected continuum of Nash equilibria, there might be no $\sigma \in \mathcal{N}((\pi_p)_p)$ for which there is a maximal-expectations subsequence (I do not know of any example of this for sophisticated Bayesian learning, although such examples are easily constructed if the assumption that each μ_p is u.a.c. is dropped). However, if $(\pi_p)_p$ is "regular," $\mathcal{N}((\pi_p)_p)$ is a finite set, so $L((e_p)_p, h_\infty)$ is a finite set by Theorem 2. In this case, the history h_∞ can be partitioned into a finite number of subhistories each of which is maximal for an element of $L((e_p)_p, h_\infty)$. Roughly stated, a game is "regular" if any small perturbation of the payoff functions induces a small perturbation of the Nash equilibria. The set of nonregular games is a closed, nowhere-dense subset of G having Lebesgue measure zero. If this set is added to the set $G'((e_p)_p)$ in the statement of Theorem 3', then the pathology mentioned above is avoided.

The final result of this section concerns the rate at which expectations converge under sophisticated Bayesian learning. It is proved in Jordan (1992a) that if expectations converge to a "regular" Nash equilibrium, then convergence is exponential. The proof is quite long, so no explanation of the argument will be attempted here. The 2×2 example in Section 3 reveals the geometrical basis for the result. The set G'' in Theorem 4 is the set of nonregular games.

Definition 2.9 *A sequence* $\{\sigma_k\}_{k=1}^{\infty} \subset \Delta(S)$ *is said to converge exponentially to an element* $\sigma \in \Delta(S)$ *if there exist constants* $a > 0$ *and* $0 < \lambda < 1$ *such that* $\|\sigma_k - \sigma\| < a\lambda^k$ *for all k.*

Theorem 5 *There is a closed, nowhere-dense set* G'' *with Lebesgue measure zero such that if* $(\pi_p)_p \notin G''$ *and* $(e_p)_p$ *is a sophisticated Bayesian learning* *process, then there is a best-response history h_∞ for $(\pi_p)_p$ and $(e_p)_p$ with the property that if $\{h_{\infty|t_k}\}_{k=1}^\infty$ is a subhistory with $e_{t_k}(h_{\infty|t_k}) \to \sigma$ for some $\sigma \in \mathcal{N}((\pi_p)_p)$, then $e_{t_k}(h_{\infty|t_k})$ converges exponentially to σ.*

3. SOPHISTICATED BAYESIAN LEARNING IN 2×2 GAMES

In general, the expectation functions that constitute a sophisticated Bayesian learning process can be computationally complex, because Definition 2.6 (ii) requires the computation of a Bayesian Nash equilibrium each period. For 2×2 games, however, there is a natural choice of the prior distributions $(\mu_p)_p$, which makes SBL geometrically transparent. The following discussion extends the description given by Jordan (1991b, Section 2.7).

Let player 1 be the row player, let player 2 be the column player, and let $S_1 = \{T, B\}, S_2 = \{L, R\}$. Then each player p's payoff function is a 2×2 matrix with the entries $\pi_p(T, L), \pi_p(T, R), \pi_p(B, L), \pi_p(B, R)$. However, it will be convenient to employ the following normalization, which reduces each player p's space of possible payoff functions to the unit circle in R^2. To motivate this normalization, suppose that player 1 anticipates that player 2 will play L with probability $\sigma_2(L)$. Then player 1's optimal strategy is T, B, or both, as the quantity

$$\sigma_2(L)[\pi_1(T, L) - \pi_1(B, L)] + (1 - \sigma_2(L))[\pi_1(T, R) - \pi_1(B, R)]$$

is positive, negative, or zero, respectively. Therefore we can subtract the second row of player 1's payoff matrix from each row, so that the top row is now $(\pi_1(T, L) - \pi_1(B, L), \pi_1(T, R) - \pi_1(B, R))$, and the bottom row is $(0, 0)$. Applying the same normalization to the columns of player 2's payoff matrix reduces the payoff bimatrix to the form

	L	R
T	a, α	$b, 0$
B	$0, \beta$	$0, 0$

If we ignore the measure-zero possibility that $a = b = 0$, that is, $\pi_1(T, L) = \pi_1(T, R)$ and $\pi_1(B, L) = \pi_1(B, R)$, then we can further normalize (a, b) to the unit circle without affecting player 1's response to any mixed strategy played by player 2. Thus we can assume that $a^2 + b^2 = 1$, and for player 2, that $\alpha^2 + \beta^2 = 1$. Under this normalization, each player's payoff function is a point on the unit circle, and each 2×2 game is a point on the torus (we will continue to exclude the degenerate cases $a = b = 0$ and $\alpha = \beta = 0$).

One immediate benefit of this normalization is a simple graphical representation of the Nash equilibrium correspondence for 2×2 games. Figure 1 pictures the Nash equilibrium correspondence when points on the unit circle

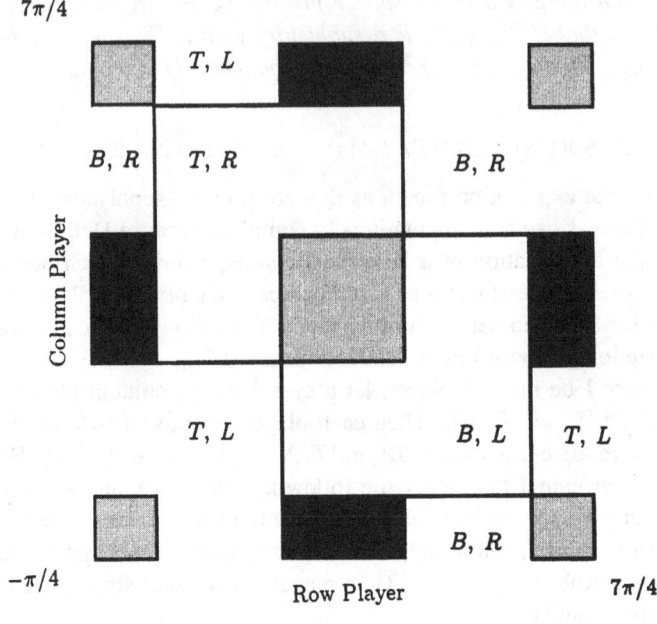

Figure 1.

are measured in radians, with $(1, 0) = 0 = 2\pi$ and $(0, 1) = \pi/2$. Thus, the game $(-\pi/4, -\pi/4)$ at the lower left corner of Figure 1 is the bimatrix game

	L	R
T	$1/\sqrt{2}, 1/\sqrt{2}$	$-1/\sqrt{2}, 0$
B	$0, -1/\sqrt{2}$	$0, 0$

Since $-\pi/4 = 7\pi/4$, the two sides of the figure are identified, and the top and bottom of the figure are identified, so Figure 1 is the usual square depiction of the torus (e.g., Guillemin and Pollak [1974], Figure 1-10, p. 17). In the white areas, the Nash equilibrium is a unique pure-strategy equilibrium, so each white area is labelled with the respective equilibrium pure strategies. The gray areas are intersections of white areas. Therefore, each game in the gray square $[\pi/2, \pi] \times [\pi/2, \pi]$, in the center of Figure 1, for example, has the two pure-strategy equilibria (T, R) and (B, L). Games in the interior, $(\pi/2, \pi) \times (\pi/2, \pi)$, also have a single mixed-strategy equilibrium. These games behave as "battle of the sexes" games. Note that because of the identifications along the sides and the top and bottom of Figure 1, the four small, gray squares in the corners adjoin to form the square $[3\pi/2, 0] \times [3\pi/2, 0]$

164

on the torus. Each game in the interior of a black area has a mixed-strategy equilibrium as its unique Nash equilibrium. The games on the boundaries of regions are "singular games," which typically have a continuum of Nash equilibria.

We first need to specify a prior probability distribution over each player's space of possible payoff functions, in this case, over the unit circle for each player. A natural choice is the uniform distribution for each player. For this prior distribution, we will study SBL along the two-period history $((T, R), (B, L))$. More precisely, we will compute the first-period expectations $e_0(\cdot)$ and second-period expectations $e_1(\cdot|T, L)$, which are uniquely determined, and show that there are three possible choices for the third-period expectations $e_2(\cdot|(T, L), (B, R))$. First, let σ_2 denote the initial probability distribution on $S_2 = \{L, R\}$ facing player 1. That is, $\sigma_2(L)$ is the probability that player 2 will play L in the first period. Then player 1 will play T in period 1 if player 1's payoff function, represented by (a, b), satisfies $a\sigma_2(L) + b(1 - \sigma_2(L)) > 0$. That is, the set of player 1 "types" that play T in period 1 is the set $T_1 = \{(a, b) : a\sigma_2(L) + b(1 - \sigma_2(L)) > 0\}$. The set $\{(a, b) : a\sigma_2(L) + b(1 - \sigma_2(L)) = 0\}$ has prior probability zero and thus can be ignored. Hence, for any expectation $\sigma_2(L)$, the probability that player 1 will play T, which is simply the prior probability of the set T_1, equals 1/2. Since this reasoning applies to both players symmetrically, the unique first-period Bayesian Nash equilibrium expectations are $e_{20}(T) = e_{20}(B) = 1/2$ and $e_{10}(L) = e_{10}(R) = 1/2$. Figure 2 illustrates this reasoning for player 1. The circle is the space of payoff functions for player 1, and the unit simplex represents the possible first-period expectations $(e_{10}(L), e_{10}(R))$. The semicircle above the line perpendicular to the expectation vector is the set T_1 of player 1 "types" that play T in period 1.

Now suppose that the strategies (T, R) are played in period 1. This reveals that player 1's payoff function lies on the upper semicircle depicted in Figure 3, and that player 2's payoff function lies on the lower semicircle depicted in Figure 4. To solve for the Bayesian Nash equilibrium conditional expectations $e_1(\cdot|(T, R))$, let $x = e_{21}(T|(T, R))$, the probability that player 1 plays T in period 2, and let $y = e_{11}(L|(T, R))$. Given y, x is simply $\mu_1(\{(a, b) \in T_1 : ay + b(1 - y) > 0\})/\mu_1(T_1)$, where μ_1 denotes the uniform distribution on the unit circle. Thus x is simply the relative arc length given by the formula

$$(3.2) \qquad x = (\pi - |\theta(y)|)/\pi,$$

where $\theta(y)$ is the angle between the expectations vectors $(1/2, 1/2)$ and $(y, 1 - y)$, measured in radians. This is depicted in Figure 3. Figure 4 is the analogous diagram for player 2, and the analogous formula for y as a function of

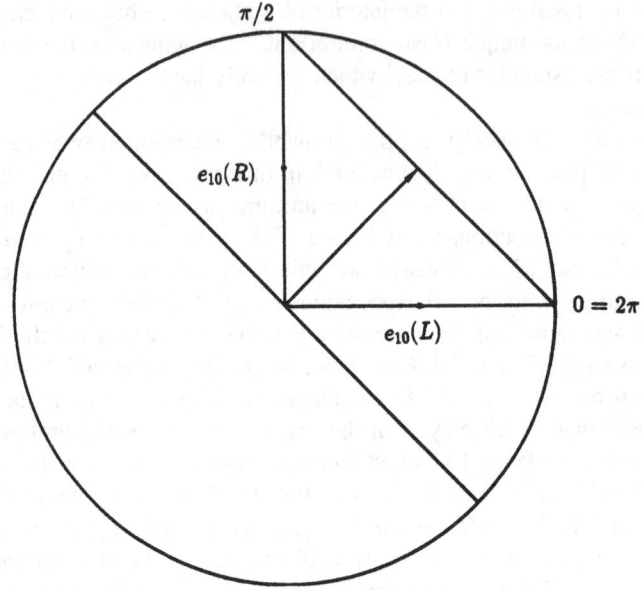

Figure 2.

x is

(3.3)
$$1 - y = (\pi - |\theta(x)|)/\pi.$$

The equations (1) and (2) have a unique solution:

(3.4)
$$x^* \approx 0.82, \quad y^* \approx 0.18,$$

so the unique Bayesian Nash equilibrium expectations are $e_{21}(T|(T, R)) = x^*$ and $e_{11}(L|(T, R)) = y^*$.

Now suppose that the strategies (B, L) are played in period 2. This reveals that player 1's payoff function lies in the arc between the line perpendicular to the expectation vector $(1/2, 1/2)$ and the line perpendicular to the expectation vector $(y^*, 1-y^*)$, and the analogous inference can be drawn concerning player 2. Algebraically, (a, b) and (α, β) have been revealed to satisfy

(3.5)
$$(1/2)a + (1/2)b > 0,$$
$$y^*a + (1 - y^*)b < 0;$$
$$(1/2)\alpha + (1/2)\beta < 0; \text{ and}$$
$$x^*\alpha + (1 - x^*)\beta > 0.$$

166

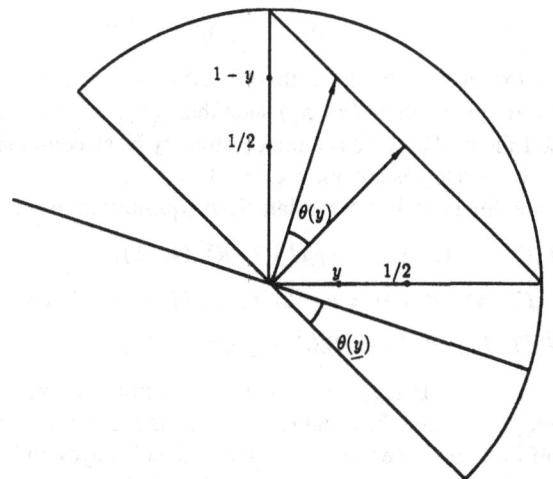

Figure 3.

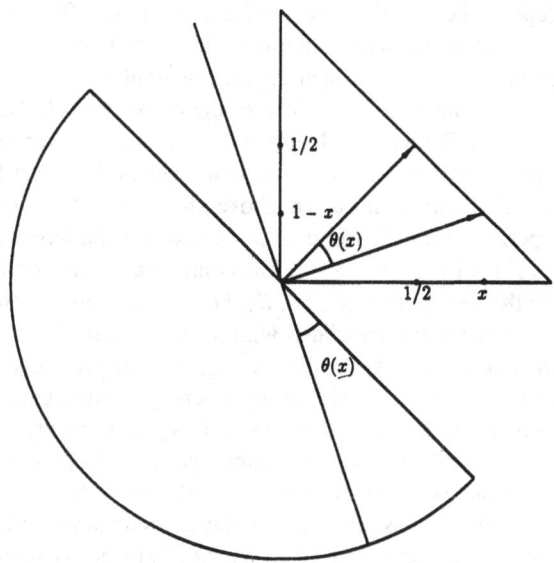

Figure 4.

167

Since $y^* < 1/2$ and $x^* > 1/2$, (4) implies

$$(3.6) \qquad a > 0, \quad b < 0; \quad \alpha > 0, \quad \beta < 0.$$

It follows from (5), or from Figure 1, that (T, L) and (B, R) are pure-strategy Nash equilibria for every game (π_1, π_2) such that $((\pi_1, \pi_2), (T, R), (B, L)) \in$ supp φ. It also follows that there is a mixed-strategy Nash equilibrium, but the equilibrium mixed strategies are not yet revealed.

Thus there are three possible Bayesian Nash equilibrium expectations:

$$(3.7) \quad e_{22}(T|(T, R), (B, L)) = e_{12}(L|(T, R), (B, L)) = 1;$$

$$e_{22}(T|(T, R), (B, L)) = e_{12}(L|(T, R), (B, L)) = 0; \text{ and}$$

$$e_{22}(T|(T, R), (B, L)) \approx 0.61, e_{12}(L|(T, R), (B, L)) \approx 0.39.$$

The first two are "pure-strategy" Bayesian Nash equilibria that correspond to the pure-strategy Nash equilibria and reveal no further information about each player's payoff function. The third is a "quasi-mixed" Bayesian Nash equilibrium that further partitions the two arcs revealed by the history $((T, R), (B, L))$. This multiplicity of Bayesian Nash equilibria continues for all future periods. If the "quasi-mixed" equilibrium is selected infinitely often, it will converge to the mixed-strategy Nash equilibrium determined by the true payoff functions $(a, b), (\alpha, \beta)$.

Figure 5 depicts the possible limits of expectations. The white, black, and gray regions have the same interpretation as in Figure 1. Where the Nash equilibrium is unique, it is also the unique limit of any Bayesian Nash equilibrium-expectations sequence. Where there are multiple Nash equilibria, a comparison of Figure 1 and Figure 5 indicates that for some games the expectations limit is unique, and for some games there can be multiple cluster points. The games discussed above, for which (T, R) and (B, L) are played in periods 1 and 2, respectively, constitute the large, gray square $\approx [-\pi/4, -\pi/15] \times [47\pi/30, 7\pi/4]$. The contiguous, smaller gray square is associated with the history $((T, R), (T, R), (B, L), \ldots)$, and so on, with each successively smaller square associated with an additional play of (T, R) followed by a switch to (B, L). As the picture suggests, the gray squares decrease exponentially to the point $(0, 3\pi/2)$. More generally, all four cascades of gray squares correspond to successively longer runs of a given strategy pair followed by a simultaneous switch to the opposite strategies by both players.

In general, there are several distinct possible phases of the learning process. The initial phase continues as long as both players maintain their initial strategies. During this phase expectations converge rapidly to the discrete distribution at the initial strategies. For an initial run of (T, L), for example,

$$e_{11}(T, L)(L) \approx 0.8192$$

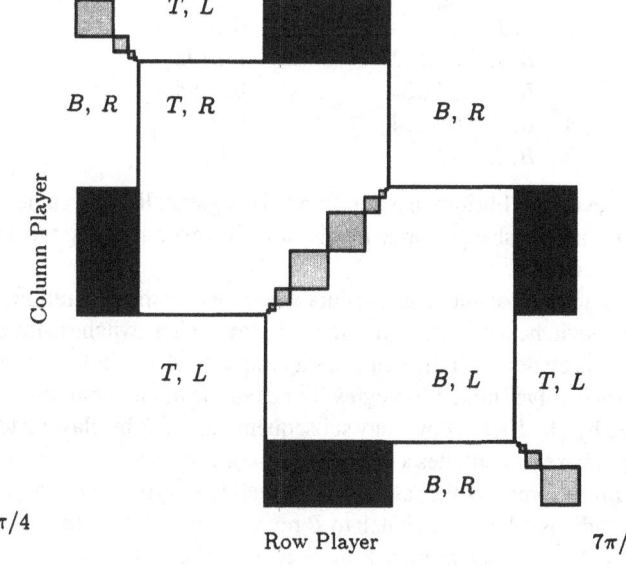

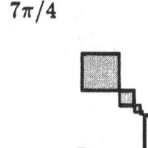

Figure 5.

$$e_{12}((T, L), (T, L))(L) \approx 0.9402$$
$$e_{13}((T, L), (T, L), (T, L))(L) \approx 0.9815; \text{ etc.}$$

If both players first change strategies at the same iteration, learning enters and remains in the multiple-equilibrium phase described above. Alternatively, if one player changes strategies before the other, a different phase is entered in which expectations are always uniquely determined. For example, if the first departure from (T, L) is (B, L), $e_{2t}(\cdot)(T)$ drops abruptly and begins to approach zero even though player 1 may continue to alternate between T and B. This phase continues as long as player 2 continues to play L. Typically, after a finite number of periods, expectations jump to the discrete equilibrium $\sigma(B, L) = 1$.

For example, the game

	L	R
T	$-2, -1$	$3, 0$
B	$0, 10$	$0, 0$

169

has the sequence

t	s_t	$e_{1(t-1)}(h_{\infty|t-1})(L)$	$e_{2(t-1)}(h_{\infty|t-1})(T)$
1	T, L	0.5	0.5
2	B, L	0.8192	0.8192
3	B, L	0.7624	0.1494
4	B, L	0.7624	0.0
5	B, L	1.0	0.0

which achieves equilibrium in 5 iterations. This game, like the game discussed earlier, has three Nash equilibria. In this case, however, learning process makes a unique selection.

The remaining possible phase occurs when, one or more iterations after the first player switches strategies, the second player also switches strategies. At that point it is evident that the game has a unique Nash equilibrium and that the equilibrium involves mixed strategies. For example, if an initial run of (T, L) is terminated by (B, L), as above, any subsequent play of R by player 2 terminates the second phase and initiates a final phase of convergence to the mixed-strategy equilibrium. Expressing the respective normalized payoff functions, (a, b) and (α, β), in radians, player 1's switch to B reveals that $(\pi/2) < (a, b) < (3\pi/4)$, and player 2's subsequent switch to R reveals that $-(\pi/4) < (\alpha, \beta) < 0$. Hence the game is located in the black rectangle at the bottom of Figure 1.

The mixed-strategy phase provides the clearest demonstration of exponential convergence. Consider the game

	L	R
T	$-2, 3$	$3, 0$
B	$0, -2$	$0, 0$

which generates the plays (T, L), (B, L), (B, R) in the first three periods. After period 3, the payoff functions of the two players are revealed to lie in the respective arcs shown in Figure 6.

In each subsequent period t, player 1's remaining arc will be split into two pieces, corresponding to the "types" who will play T in period t and those who will play B. If B is played in period t, the T-types are discarded and the remaining arc is again split in period $t + 1$. The Bayesian Nash equilibrium determination of expectations following period t forces the relative measure of T-types in period $t + 1$ to equal $e_{2t}(h_t)(T)$, which, by Theorem 2, is converging to the Nash equilibrium $0 < \sigma_1(T) < 1$. Hence player 1's arc length shrinks exponentially to zero. Since player 2's arc is also shrinking exponentially, expectations must converge exponentially to the unique Nash equilibrium. Table 1 illustrates this process for the game defined above. The Nash equilibrium for this game has $\sigma_1(T) = 0.4, \sigma_2(L) = 0.6$. The last two columns, δ_1 and δ_2,

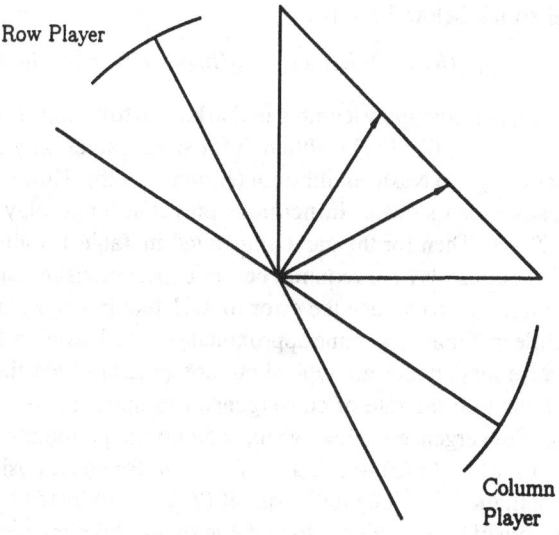

Row Player

Column
Player

Figure 6.

Table 1.

t	s_t	$e_{1(t-1)}(h_{\infty|t-1})(L)$	$e_{2(t-1)}(h_{\infty|t-1})(T)$	δ_1	δ_2
1	T, L	0.5	0.5	2.0	2.0
2	B, L	0.8192	0.8192	1.0	1.0
3	B, R	0.7624	0.1494	0.1808	0.8192
4	B, R	0.6514	0.3917	0.1538	0.1946
5	T, L	0.5808	0.4552	0.0936	0.0679
6	B, L	0.6203	0.4314	0.043	0.0394
7	B, L	0.6037	0.4158	0.0242	0.0245
8	T, L	0.5944	0.4061	0.0141	0.0148
9	T, L	0.6000	0.4004	0.0057	0.0088
10	B, R	0.6022	0.3970	0.0023	0.0053
11	B, R	0.6013	0.3990	0.0014	0.0021
12	B, R	0.6007	0.3998	0.0008	0.0008

give the arc lengths, in (radians/π), for the respective players at the beginning of each period.

Exponential convergence can also be demonstrated when expectations converge to pure-strategy equilibria, but the reasoning is much less transparent. For example, for games that have the constant history (T, L), the player arc lengths never shrink below $3\pi/4$ but

$$\lim_{t\to\infty}(1 - e_{1(t+1)}(h_{\infty|t+1})(L))/(1 - e_{1t}(h_{\infty|t})(L)) = 3\pi/(4 + 3\pi).$$

For 2×2 games, fictitious-play learning is also known to converge to Nash equilibrium (e.g., Rosenmüller 1971), although for some games only the marginal frequencies converge to Nash equilibrium (Jordan 1992b). However, fictitious play is typically much slower. In fictitious play, the initial play is arbitrary, so let $s_1 = (T, L)$. Then for the game illustrated in Table 1, more than 7,600 iterations of fictitious play are required before expectations remain within .01 of Nash equilibrium. To reduce the error to .001 requires more than 680,000 iterations, while in Table 1 the same approximation is achieved in 12 iterations.

These relative magnitudes are typical but not general. Over the torus there is no lower bound on the rate of convergence (no upper bound on λ in Definitions 2.9). Convergence is slow when, modulo the permutation of players or strategies, $(a, b) = (\pi/2) + \varepsilon$ and $(\alpha, \beta) = -\delta$ for small positive ε and δ. When ε is small, there is a long initial run of (T, L), terminated by (B, L) and followed by a run of $(\cdot, L)$ which is long if δ is small. After the second phase is terminated by $(\cdot, R)$, expectations converge to the mixed-strategy equilibrium at a slow exponential rate. Of course ε and δ are forced to zero exponentially in the length of the first and second phase, respectively, so, stated very loosely, the measure of slow games is small.

The 2×2 example of sophisticated Bayesian learning (SBL) illustrates several respects in which SBL differs sharply from fictitious play and other learning processes motivated by statistical-estimation methods. In addition to converging more rapidly, SBL may exhibit discontinuous phase transitions. In many cases, the players may jump to an exact pure-strategy equilibrium after a small number of iterations, even though initial expectations are evenly mixed. Finally, the observation of a particular strategy in period t need not increase, and may decrease, the expected probability of observing that strategy in period $t + 1$.

4. CONCLUSION

Perhaps the clearest way to interpret sophisticated Bayesian learning is to think of it as a software package. Imagine that a software company has produced for sale a program that computes (or reads from a table) the expectations $(e_{-pt}(\cdot))_{p,t}$ determined by uniform priors $(\mu_p)_p$. In order for you, as a player, to use this program, you must enter the number of players, the cardinality of each player's strategy set S_p, and your own player index, p. Then, after each period t, you enter the strategy n-tuple $s_t = (s_{1t}, \dots, s_{nt})$ that was played in period t, and the program responds by giving you the expectation $e_{-pt}(s_1, \dots, s_t) \in \Delta(S_{-p})$.

You then consult your payoff function π_p to determine a best response s_p, which you play in period $t + 1$. There is a "deluxe" version of the program that will compute a best response s_p if you enter π_p. If all players use this program, then for every game $(\pi_p)_p$, with the possible exception of a set of Lebesgue measure zero, expectations will asymptotically lie in the set $\mathcal{N}((\pi_p)_p)$ of Nash equilibria. If all players use the deluxe version, then the program will select appropriately when there are multiple best responses so that convergence will obtain for all games. It may be worth emphasizing that players can use the program independently, that is, no network is needed.

The software analogy illustrates both the principal strength and the principal shortcoming of sophisticated Bayesian learning. On the one hand, SBL leads to Nash expectations for all games. In this sense, SBL provides a "foundation" for Nash equilibrium. On the other hand, it requires all players to use the same method of expectation formation. Put somewhat differently, payoff characteristics are completely decentralized, but expectation formation is completely centralized.

Of course, most learning models constructed to date impose considerable if not complete uniformity of learning mechanisms across players. The two other Bayesian learning models presented to this conference are less restrictive than most in this respect. Kalai and Lehrer (1991a) allow the complete decentralization of prior beliefs over histories of play, and thus complete decentralization of expectation formation. However, convergence is obtained only when the true best-response distribution is absolutely continuous with respect to each player's prior. Nyarko (1991) allows complete decentralization of prior beliefs over payoff functions, but the expectation functions are then determined, as in the present paper, via Bayesian Nash equilibrium. The maximal decentralization of both payoff characteristics and expectation formation consistent with convergence to Nash equilibrium remains one of the major unknowns in the theory of learning in games.

REFERENCES

Blume, L., and D. Easley. 1992. "What Has the Rational Learning Literature Taught Us?" Mimeo.
Breiman, L. 1968. *Probability*. Reading, Mass.: Addison-Wesley.
Diaconis, P., and D. Freedman. 1986. "Rejoinder." *The Annals of Statistics* **14**:63–64.
Guillemin, V., and A. Pollak. 1974. *Differential Topology*. Englewood Cliffs, N.J.: Prentice-Hall.
Jordan, J. 1991a. "Bayesian Learning in Normal Form Games." *Games and Economic Behavior* **3**:60–81.
Jordan, J. 1991b. "Bayesian Learning in Repeated Games." Mimeo.
Jordan, J. 1992a. "The Exponential Convergence of Bayesian Learning in Normal Form Games." *Games and Economic Behavior* **4**:202–17.
Jordan, J. 1992b. "Three Problems in Learning Mixed-Strategy Nash Equilibria." Mimeo.
Kalai, E., and E. Lehrer. 1991a. "Rational Learning Leads to Nash Equilibria." Mimeo.
Kalai, E., and E. Lehrer. 1991b. "Subjective Equilibrium in Repeated Games." Mimeo.

Nyarko, Y. 1991. "Bayesian Learning without Common Priors and Convergence to Nash Equilibrium." Mimeo.

Nyarko, Y. 1992. "Bayesian Learning in Repeated Games Leads to Correlated Equilibria." Mimeo.

Rosenmüller, J. 1971. "Über Perioizitätseigenschaften spieltheoretischer Lernprozesse." *Zeitschrift für Wahrscheinlichkeitstheorie und verwandte Gebeite* **17**:259–308.

Shapley, L. 1964. "Some Topics in Two-Person Games." In Dresher, M. et al., eds., *Advances in Game Theory*. Annals of Mathematical Studies **52**:1–28.

Young, P. 1991. "The Evolution of Conventions." Working paper no. 91-10-043, SFI Economics Research Program, Santa Fe Institute.

9

Savage–Bayesian agents play a repeated game

YAW NYARKO

Abstract

I study the limiting behavior of an infinitely repeated game in which the players have imperfect information about the strategies and types of the other players in the game. The players are assumed to be "Savage–Bayesian" agents. In particular, the players form prior probability beliefs over the strategies and types of the other players in the game and maximize subjective expected utility. I discuss the different assumptions that may be placed on the priors of such players. I state the convergence results that are possible under the various assumptions on the priors. I focus on the question of when the limiting behavior is a Nash equilibrium of the true game. I illustrate the possible results with examples and counterexamples.

1. INTRODUCTION

I study the play of an infinitely repeated game by a finite collection of players. I suppose that the players are "Savage–Bayesian" agents. In particular, their preferences obey the axioms of Savage (1954), and hence they have beliefs over all items for which they have imperfect information, and then they maximize (subjective) expected utility given these beliefs. Such players are therefore necessarily "Bayesian" in the sense that they use Bayes's rule (i.e., conditional expectations and the laws of probability) in updating beliefs after receiving information during the course of the play of the game. Such "Savage–Bayesian" agents are just like agents in most single-agent decision theory models of economics: they maximize expected utility given their beliefs.

What is the long-run behavior of the game played by such Savage–Bayesian agents? When may we conclude that in the long run the game looks like the outcome of a game in a Nash equilibrium? Suppose that under fairly general conditions we are able to obtain limiting behavior that looks like a

I am grateful to C. V. Starr Center, the Challenge Fund, and the Presidential Fellowship at New York University for their generosity.

Nash equilibrium. Then we would have a justification for the use of the Nash equilibrium concept in game theory and economics via "Bayesian learning." We could then say that players "learn their way" to a Nash equilibrium. If, alternatively, we are unable to obtain such limiting results, then our use of the Nash equilibrium concept is called into question. We should then either seek other explanations for their use or adopt broader modelling constructions in economics and game theory. Hence this paper asks the question posed much earlier in Blume and Easley (1984).

One may object to the use of "Savage–Bayesian" agents in studying the question of whether limiting behavior looks like a Nash equilibrium. Some often-heard criticisms are that, on the one hand, such agents are presupposed to have "too much knowledge" about the structure of the game and, on the other hand, that they are able to form beliefs that are just too complicated for the human mind to have. Such criticisms of course are nonsense and completely miss the point. The assumption that a player is a Savage–Bayesian agent merely means that the agent forms some beliefs and then chooses optimal strategies given those beliefs. Nowhere is it being claimed that such agents have "correct" beliefs in any sense of the word. If, for example, a player believes that the opponent will choose an action a_1 in each period even though in each of the previous 50 billion periods the action a_2 has been chosen, this does not contradict the fact that the player is a "Savage–Bayesian" agent. If the agent has a "simple" or "naive" rule of thumb for predicting how the opponent will play in the future given any past history, this, too, does not contradict the fact that the player is "Savage–Bayesian," even when such rules are oversimplifications of reality. I formalize this remark in section 7, entitled "We Are All Savage–Bayesians." (This point has also been made formally in Jordan (1992). See also the discussion in Kiefer and Nyarko (1993).)

The relevant issue is not whether or not players are "Savage–Bayesian," since almost everything can be rationalized as being "Savage–Bayesian." The issue is whether the assumptions placed on the beliefs of players are "good" assumptions or not. In this paper I discuss some of the various assumptions that may be placed on the priors of the players. There are really two classes of assumptions one may place on the priors. The first class are those that are placed on each individual prior of an agent. I discuss this class in section 8. This class of assumptions does not usually generate much controversy in the literature. The second class of assumptions are those that say how the beliefs of one player should be related to the beliefs of another player. I discuss this class in section 9. Most of the debate is typically over the second class of assumptions. (For example, most of the criticisms of the Savage–Bayesian "agents" are actually criticisms of the common prior assumption requiring that agents have ex ante priors that are the same.)

The different assumptions on the priors yield different conclusions to the question of the limiting behavior of the game. I provide in section 10 a very quick and informal list of the different conclusions that result from the different assumptions. These conclusions are then illustrated in fairly self-contained examples in section 11. In section 12, after developing the required notation, I state formally the conclusions that were informally listed in section 10 and illustrated in section 11. One should consult Blume and Easley (1992) for a survey of some of the results in the rational-learning literature with a different emphasis.

2. SOME TERMINOLOGY

I is the *finite* set of players. Given any collection of sets $\{X_i\}_{i \in I}$, we define $X \equiv \prod_{i \in I} X_i$ and $X_{-i} \equiv \prod_{j \neq i} X_j$. Given any collection of functions $f_i : X_i \to Y_i$ for $i \in I$, $f_{-i} : X_{-i} \to Y_{-i}$ is defined by $f_{-i}(x_{-i}) \equiv \prod_{j \neq i} f_j(x_j)$. The Cartesian product of metric spaces will always be endowed with the product topology. Let X be any metric space. We let $\mathcal{P}(X)$ denote the set of (Borel) probability measures on X. Unless otherwise stated the set $\mathcal{P}(X)$ will be endowed with the weak topology. Given any $\sigma \in \mathcal{P}(X)$ we let $\sigma(dx)$ denote integration: $\int h(x)\sigma(dx)$ is the integral of the real-valued function h on X with respect to σ. If X is a Cartesian product $X = YZ$, we let $\sigma(dy)$ denote integration over Y with respect to the marginal of σ on Y. The latter will often be denoted by $\mathrm{Marg}_Y \sigma$ or simply as $\sigma(dy)$. $\mathcal{R}$ denotes the real line. Suppose for $k = 1, 2, \ldots K$, ν_k is a probability measure on a complete and separable metric space X_k. Then we let $\nu \equiv \prod_{k=1}^{K} \nu_k$ denote the product of the measures $\{\nu_k\}_{k=1}^{K}$ on the Cartesian product $\nu \equiv \prod_{k=1}^{K} X_k$.

3. THE PRIMITIVES OF THE MODEL

3.1. Recall that I is the *finite* set of players. S_i represents the *finite* set of actions available to player i at each date $n = 1, 2, \ldots$; we define $S \equiv \prod_{i \in I} S_i$. Even though the action space S_i is independent of the date, I will sometimes write S_i as S_{in} when I seek to emphasize the set of action choices at *date n*. I define $S^N \equiv \prod_{n=1}^{N} S_n$ and $S^\infty \equiv \prod_{n=1}^{\infty} S_n$, the set of data n and infinite histories, respectively. s^0 will denote the null history (at date 0, when there is no history). In summary, s or S with a superscript (e.g., s^N) denotes the history, while with a subscript (e.g., s_n) denotes the current period. *Perfect recall* is assumed; in particular, at date n when choosing the date n action S_{in}, player i will have information on $s^{N-1} = \{s_1, \ldots, s_{N-1}\}$. I define $F_{iN} \equiv \{f_{iN} : S^{N-1} \to \mathcal{P}(S_i)\}$; $F_N \equiv \prod_{i \in I} F_{iN}$; $F \equiv \prod_{N=1}^{\infty} F_N$; $F_i \equiv \prod_{N=1}^{\infty} F_{iN}$. F_i is the set of all *behavior strategies* for player i. F_{iN} is endowed with the topology of pointwise convergence; F_N, F_i, and F are endowed with their respective product topologies. The mapping $m:F \rightarrow \mathcal{P}(S^\infty)$ defines the probability distribution $m(f)$ on S^∞ resulting from the behavior strategy profile f. In particular, it is the distribution induced by the following transition equation:

$$(3.1.1) \qquad \forall D \subseteq S_{N+1}, \qquad m(f)(D \mid s^N) \equiv f_{N+1}(S^N)(D)$$

3.2. The type space. Each player $i \in I$ has an attribute vector that is some element θ_i of the set Θ_i. The attribute vector will represent the parameter of player i's utility function, which is unknown to other players in the game. The function (or mapping) $u_i:\Theta_i \times S \rightarrow \mathcal{R}$ is player i's (within period or instantaneous) utility function, which depends upon her attribute vector, θ_i, as well as the vector of actions, $s \in S$, chosen by the players. I assume that u_i is *continuous and bounded* on its domain. I will suppose that Θ_i is a *compact* subset of finite-dimensional Euclidean space. This is without loss of generality, since the set of joint actions, S, is assumed finite. Player i has a discount factor that is a *continuous* function, $\delta_i : \Theta_i \rightarrow (0, 1)$, of the player i's attribute vector.

Suppose that players know the functional forms of each player's utility function, $\{u_i\}_{i \in I}$. Each player i knows her own attribute vector θ_i but does not know those of other players, θ_{-i}. Player i's type, τ_i, specifies that player's attribute vector, θ_i. Let T_i denote the set of possible types of player i, and set $T \equiv \prod_{i \in I} T_i$. Let $\theta_i(\tau_i)$ denote the attribute vector of player i of type τ_i; assume that this is *continuous*[1] in θ_i. The type space will be assumed to be a complete and separable metric space.

3.3. Payoffs. We define $U_i : \Theta_i \times S^\infty \rightarrow \mathcal{R}$ and $V_i : \theta_i \times F \rightarrow \mathcal{R}$ by

$$U_i(\theta_i, S^\infty) = \sum_{n=1}^{\infty} [\delta_i(\theta_i)]^{n-1} u_i(\theta_i, S_n) \quad \text{and}$$

$$V_i(\theta_i, f) = \int U_i(\theta_i, S^\infty) m(f)(ds^\infty),$$

where $s^\infty \equiv \{S_n\}_{n=1}^{\infty}$ and where $m(f)(ds^\infty)$ denotes integration with respect to the measure $m(f)$ over S^∞ (as defined in (3.1.1)).

4. NASH EQUILIBRIUM CONCEPTS

Consider as *fixed* an element $\theta \equiv \{\theta_i\}_{i \in I}$ of the space of attribute vectors, Θ. Define for each $i \in I$

$$N_i(\theta_i) \equiv \left\{ f = \prod_{j \in I} f_j \in F : f_i \in \arg\max V_i(\theta_i, f_{-i}, \cdot) \right\};$$

178

$$N(\theta) \equiv \cap_{i \in I} N_i(\theta_i); \text{ and}$$

$$ND(\theta) \equiv \{v \in \mathcal{P}(S^\infty) | \exists f \in N(\Theta) \text{ with } v = m(f),$$

$$\text{where } m(f) \text{ is as in } (3.1.1)\}.$$

$N(\theta)$ is the set of Nash equilibrium *behavior-strategy* profiles for the game with attribute vector θ. $ND(\theta)$ is the set of all Nash equilibrium *distributions*—the distributions of play that can be generated by some Nash equilibrium behavior-strategy profile. I emphasize that each of these Nash equilibrium concepts are defined for a *fixed-attribute* vector $\theta = \{\theta_i\}_{i \in I}$.

5. THE BELIEFS OF THE SAVAGE–BAYESIAN PLAYERS

5.1. I will posit the existence of probability measures, μ_i for each $i \in I$, over the set of all possible vectors of types, behavior strategies, and actions, $T \times F \times S^\infty$. μ_i will represent the prior belief of player i. The interpretation shall be that for each type τ_i of player i, the conditional probability, $\mu_i(. \mid \tau_i)$, will represent the belief held by player type τ_i. I will refer to μ_i as an *ex ante subjective belief* for player i if μ_i is a probability measure over $T \times F \times S^\infty$ that "respects" the function $m(f)$ in (3.1.1); that is, the function $m(f)$ in (3.1.1) is a version of the conditional probability over S^∞ of μ_i given (τ, f) so that we may write $\mu_i(ds^\infty \mid \tau, f) = m(f)(ds^\infty)$ for μ_i-a.e. $(\tau, f) \in T \times F$. In the rest of this section, I provide further motivation for these beliefs by sketching how they may be "constructed." (See Nyarko (1993b) for formal details.)

5.2. Construction of the beliefs. Each player i of type i will enter the game with some belief about the play and the types of the other players in the game. Each player type will be assumed to know her own type, τ_i, and the (possibly randomized) behavior strategy she chooses. Hence each player type τ_i may be assumed to enter the game with a belief about the types, behavior strategies, and actions of all players (including player i herself). We denote this by $\mu_i(.; \tau_i)$; in particular, $\mu_i(.; \tau_i)$ is a probability measure over $T \times F \times S^\infty$ that assigns probability one to some $\tau_i \in T_i$.

I now suppose that there is an "ex ante" or "behind-the-veil" period. During this period, "nature" chooses a vector of types for the players. I suppose that each player has some ex ante belief about her type, $\tau_i \in T_i$, which I denote by $\hat{\mu}_i$. Combining this with the conditional probability $\mu_i(.; \tau_i)$ just "constructed" results in a joint distribution μ_i over $T \times F \times S^\infty$ such that (i) the marginal of μ_i over T_i is equal to $\hat{\mu}_i$ and (ii) (a version of) the conditional probability of μ_i conditional on τ_i is equal to $\mu_i(.; \tau_i)$. The precise nature of this ex ante probability plays no role in the decision making of player i. Player i chooses a behavior strategy *after* she has observed her type τ_i, at which time she will have beliefs defined via $\mu_i(.; \tau_i)$, regardless of the marginal on τ_i. The use of the ex ante distribution $\hat{\mu}_i$ allows us to avoid measurability problems. Further, many results will not be true for all types τ_i but instead on a set of types "with probability one"; the use of the ex ante allows us to make the latter statements and hence allows us to ignore small sets of troublesome and uninteresting values of types. All the details are in Nyarko (1993b).

6. THE TRUE DISTRIBUTION

So what indeed is the true distribution describing the interaction of the players? Well, first we need to specify how the types of players are chosen. Let us suppose they are chosen according to the distribution $\hat{\mu}^{**} \in \mathcal{P}(T)$. Fix any collection of ex ante subjective beliefs of players, $\{\mu_i\}_{i \in I}$, each of which is a probability over $T \times F \times S^\infty$. Let $\mu_i(df_i \mid \tau_i)$ be (any fixed version of) the conditional probability of μ_i over F_i given τ_i. Then $\mu_i(df_i \mid \tau_i)$ is the "true" behavior-strategy choice rule of player i *as a function* of player i's type. For any given vector of types $\tau = \{\tau_i\}_{i \in I}$, let

$$(6.1.1) \qquad \mu^{**}(df; \tau) \equiv \prod_{i \in I} \mu_i(df_i \mid \tau_i)$$

denote the product of the measures $\{\mu_i(df_i \mid \tau_i)\}_{i \in I}$ on the space of behavior strategies $F = \prod_{i \in I} F_i$. Then $\mu^{**}(df; \tau)$ is the "true" distribution of behavior strategies as a function of the types of players. Define μ^{**} to be the probability over $T \times F \times S^\infty$ such that

  (i) μ^{**} has a marginal over T equal to $\hat{\mu}^{**}$;
 (ii) (6.1.1) is a version of the conditional distribution of μ^{**} over F conditional on $\tau \epsilon T$; and
(iii) $\mu^{**}(ds^\infty \mid \tau, f) \equiv m(f)(ds^\infty)$ is a version of the conditional of μ^{**} over S^∞ given (τ, f) where $m(f)$ is as in (3.1.1).

More formally, the measure μ^{**} is defined by setting for each continuous and bounded real-valued function H on $T \times F \times S^\infty$, $\int H(\tau, f, s^\infty)d\mu^{**} = \int_T \int_F [\int H(\tau, f, s^\infty)m(f)(ds^\infty)]\mu^{**}(df; \tau)\hat{\mu}^{**}(d\tau)$. The measure μ^{**} defined above is the "true" ex ante probability distribution when ex ante types have distribution $\hat{\mu}^{**}$ and players' ex ante subjective beliefs are given by $\{\mu_i\}^{i \in I}$. If μ^{**} is the "true" *ex ante* distribution, then we may refer to the conditional $\mu^{**}(. \mid \tau)$ (which agrees with (6.1.1) over F) as the "true" ex post distribution, obtained after the realization of types.

7. WE ARE ALL SAVAGE–BAYESIANS!

At this stage of the analysis, I have characterized each player by a belief, μ_i, which is a probability distribution over $T \times F \times S^\infty$. So far I have not placed any assumptions on these beliefs. I proceed to state formally that the assumption that players have such beliefs is itself without loss of generality.

A date N forecast function is any function that states a player's belief about the actions the players will choose after any date N history as a function of their types. More formally a date N forecast function is any mapping $e_N : T \times S^N \to \mathcal{P}(S_{n+1})$ such that $e_N(\tau, s^N)$ is a product measure over $\prod_{j \in I} S_{j,n+1}$. The date 0 forecast function is a mapping from the null history, so e_0 may be identified with an element of $\mathcal{P}(S_1)$. A forecast function is a collection of date N forecast functions $e = \{e_N\}_{N=1}^\infty$. A forecast function tells us what a player believes about the actions each type of player will play at each date as a function of the history at that date. Of course a prior μ_i defines a forecast function by setting the date N forecast to be equal to the date N conditional expectation: $e_N(\tau, s^N) = \mu_i(ds_{n+1} \mid \tau, s^N)$.

The proposition below says we may go the other way. In particular, any forecast function defines a prior μ_i over $T \times F \times S^\infty$. This therefore means that *any* rule for predicting behavior at each date, however simple or naive, may be considered to be that arising from the prior of a Savage–Bayesian agent. The imposition of a Savage–Bayesian prior is therefore without loss of generality. (A version of the result below has been stated and proved earlier by Jordan 1992.)

Proposition 7.1 *Let $e = \{e_N\}_{N=1}^\infty$ be a forecast function. Fix any ex ante distribution $\hat{\mu}$ over T. Then there exists a probability measure μ over $T \times F \times S^\infty$ such that the marginal of $\mu(. \mid \tau, s^N)$ on S_{N+1} is equal to $e_N(\tau, s^N)$ for μ-a.e. (τ, s^N) (where $\mu(. \mid \tau, s^N)$ is any fixed version of the conditional probability of μ given (τ, s^N)). Further, the measure μ will respect (3.1.1) (i.e., $\mu(ds^\infty | \tau, f) = m(f)(ds^\infty)$) and will have marginal over T equal to $\hat{\mu}$.*

PROOF: Fix any $\tau \in T$. Then for each N, the forecast function $e_N(\tau, .)$ defines a transition probability over S_{n+1} conditional on any history $s^N \in S^N$. Hence, from standard probability-extension theorems we obtain the existence of a joint probability measure over S^∞, which we denote by $\nu(\cdot; \tau)$, with conditionals defined via those transition equations; that is, such that for each $\tau \in T$,

$$\nu(ds_{N+1} | s^N; \tau) = e_N(s^N; \tau) \quad \text{for } \nu(\cdot; \tau)\text{-a.e. } s^N \in S^N \text{ and } \forall N,$$

where $\nu(\cdot \mid s^N; \tau)$ is (a version of) the conditional of $\nu(\cdot; \tau)$ given s^N. Under the independence assumption, $e(s^N; \tau)$ is a product measure on $\mathcal{P}(S_{N+1})$. We may therefore set $e(s^N; \tau) \equiv \prod_{j \in I} e_{j,N}(s^N; \tau)$ where $e_{j,N}(s^N; \tau)$ is the marginal of $e(s^N; \tau)$ on $S_{j,N+1}$. Now, $\{e_{j,N}(\cdot; \tau)\}_{N=0}^{\infty}$ defines a unique behavior strategy $f_j(\cdot; \tau) = \{f_{j,N}(\cdot; \tau)\}_{N=0}^{\infty}$ by setting $f_{j,N}(\cdot; \tau) = e_N^j(\cdot; \tau)$. It should be clear that $f(\cdot; r)$ induces a distribution over S^{∞} equal to $\nu(\cdot; \tau)$; that is, $m(f(\cdot; \tau)) = \nu(\cdot; \tau)$ where $m(\cdot)$ is as in (3.1.1).

Now define μ to be the probability over $T \times F \times S^{\infty}$ such that (i) $\mu(ds^{\infty} \mid \tau) = \nu(\cdot; \tau)$ for $\hat{\mu}$-a.e. τ (where $\mu(. \mid \tau)$ is any version of the conditional probability of μ given τ); (ii) conditional on any $\tau \in T$, μ assigns probability one to the behavior strategy profile $f(\cdot; \tau)$; and (iii) the marginal of μ over T is equal to the measure $\hat{\mu}$ given in the proposition. This measure μ can easily be seen to satisfy all the conclusions of the proposition.

8. SOME ASSUMPTIONS ON THE INDIVIDUAL PRIORS

Assumption 8.1 *The probability* $\mu_i \in \mathcal{P}(T \times F \times S^{\infty})$ *respects the probability* $m(f)$ *of (3.1.1); that is,* $\mu_i(ds^{\infty} \mid \tau, f) \equiv m(f)(ds^{\infty})$ *defines a version of the conditional probability of* μ_i *over* S^{∞} *conditional on* $(\tau, f) \in T \times F$.

Assumption 8.2 (utility maximization) $\mu_i(M_i \bigcup \hat{M}_i) = 1$ *where*

$$M_i \equiv \{(\tau, f, s^{\infty}) \in T \times F \times S^{\infty} \mid \delta_i(\theta_i) > 0 \text{ and}$$

$$f_i \text{ maximizes } \int V_i(\theta_i(\tau_i), f_{-i}, \cdot)\mu_i(df_{-i} \mid \tau_i)\} \text{ and}$$

$$\hat{M}_i \equiv \{(\tau, f, s^{\infty}) \in T \times F \times S^{\infty} \mid \delta_i(\theta_i) = 0 \text{ and}$$

$$s_{in+1} \text{ maximizes } \int u_i(\theta_i(\tau_i), s_{-in+1}, \cdot)\mu_i(ds_{-in+1} \mid s^n, \tau_i)\forall n\}.$$

Assumption 8.3 $\mu_i(df \mid \tau) = \prod_{j \in I} \mu_i(df_j \mid \tau_j)$ *for* μ_i *almost every* τ.

Assumption 8.4 *Define* $\pi_i \equiv \text{Marg}_T \mu_i$, *the marginal of* μ_i *on the type space* T_i. *Then* π_i *is a product measure on the type space* T; *that is,* $\pi_i = \prod_{j \in I}[\text{Marg}_{T_j} \pi_i]$.

Assumption 8.1 requires that μ_i respect the definition of f and in particular (3.1.1). Assumption 8.2 requires that with μ_i probability one each player i is maximizing her subjective expected discounted sum of date n utilities. Whenever the discount factor is equal to zero (i.e., on the set $\hat{M}_i$ above), player i will be required to maximize expected utility at each date. Assumption 8.2 does not imply that, under i's belief about the game, other players $j \neq i$ are maximizing their expected utility. Assumption 8.3 says that other than through their types, players have no way of correlating their actions. Assumption 8.4 is a restriction on the beliefs of players.

In my "representation" theorem of Proposition 7.1, the measure μ satisfies 8.1 by construction and satisfies 8.3 because of the independence (or product-measure) assumption implicit in the definition of a forecast function. If the ex ante distribution of types, $\hat{\mu}$, of Proposition 7.1 is a product measure over $\prod_{j \in I} T_j$, then 8.4 will hold. Many authors, Bayesian and non-Bayesian, invoke the maximizing behavior implicit in Assumption 8.2. Hence Proposition 7.1 implies that as far as assumptions 8.1–8.4 are concerned, the Savage–Bayesian hypothesis does not restrict the individual behavior of players any more than do many of the other "learning algorithms" and "rules of thumb" used in the literature.

9. "CLOSENESS" ASSUMPTIONS BETWEEN THE PRIORS

In the previous section, I discussed some assumptions that may be placed on the ex ante subjective belief, μ_i, of player i. Notice that all the assumptions of that section are about individual priors μ_i and are not assumptions that compare the prior of one agent to that of another, μ_i and μ_j for $i \neq j$. I argued in that section that those assumptions lead us nowhere, and in particular just about any behavior may be described by such priors even when one allows for maximizing behavior. To make further progress, one must impose assumptions *jointly over the priors*; that is, we must impose conditions on how μ_i is related to μ_j for all $i, j \in I$. In this section, I will discuss some assumptions of this type.

First some terminology. Given any two probability measures μ' and μ'' on some (measure) space Ω, μ' is absolutely continuous with respect to μ'' if for all (measurable) subsets D of Ω, $\mu'(D) > 0$ implies that $\mu''(D) > 0$. μ' and μ'' are mutually absolutely continuous if μ' is absolutely continuous with respect to μ'' and vice versa. We then have the following:

9.1 Common prior assumption $\mu_i = \mu_j$ *for all* $i, j \in I$.

9.2 Condition GH $\forall i, j \in I$, μ_i *and* μ_j *are mutually absolutely continuous.*

For the next two assumptions, assume the existence of some measure μ^* over $T \times F \times S^\infty$. This measure may be thought of as the true distribution of types and behavior strategies, as defined in section 6. Alternatively, we may think of this measure as that of the prior of an outside observer. This measure will be the one we would want to state all our theorems in terms of.

9.3 Condition GGH μ^* *is absolutely continuous with respect to* μ_i *for all* $i \in I$.

9.4 Condition K–L *(i) Condition (GGH) holds and (ii) there exists a countable subset* $\bar{T}$ *of the type space* T *such that* $\mu^*(\bar{T} \times F \times S^\infty) = 1$ *for all* $i \in I$.

Notice that all of these three assumptions are conditions that require that μ_i and μ_j should be "similar" or "close" in some sense. The common prior assumption is at one extreme, requiring that all the priors be equal. Condition

GH relaxes this assumption to allow the priors to be different so long as they are mutually absolutely continuous. Condition GGH requires that any event that is assigned a positive probability under μ^* should also be assigned a positive probability under μ_i. Hence if μ^* is the true distribution, condition GGH rules out the possibility that an agent assigns zero probability to an event that may occur with true positive probability. Condition GGH should be considered a generalization of condition GH. Indeed, when condition GH holds, we may set μ^* equal to the average of the μ_i's to obtain condition GGH. Alternatively, if μ^* is the true ex ante distribution, then one may show formally that condition GH implies condition GGH. Condition GGH allows for the possibility that players may entertain very diffuse beliefs about the game, but their beliefs do, however, include "the truth." The modeler, however, would want to prove theorems only in terms of the "true" distribution. Condition K–L requires in addition to condition GGH that the type space be countable. Condition GH is of course weaker than K–L. The common prior assumption and condition K–L are noncomparable, since one may hold while the other fails.

The common prior assumption is used in Jordan (1991a) and (1991b). Condition GGH is used in Nyarko (1993a) and (1993c). Kalai and Lehrer (1990) study a model in which there is only one vector of types and players form beliefs only over the strategies of others. Suppose I imbed their model into mine by supposing that there is an ex ante distribution of types. Then the mutual absolute continuity assumption used in Kalai and Lehrer (1990), after eliminating the trivial and uninteresting cases, reduces to condition K–L above. (See Nyarko 1993c for details.)

10. On the Question of Convergence to a Nash Equilibrium

10.1. From the previous discussion, we have two sets of assumptions we may place on the beliefs of players: those on individual μ_i's as in section 8, and those between the μ_i's as in section 9. As regards the assumptions of section 8, I will maintain 8.1–8.3, since I believe they are reasonable assumptions. There is, however, no reason to expect that 8.4 should necessarily hold. Hence one "parameter" in describing the possible results is whether or not 8.4 holds. Next, we have the assumptions of section 9. Conditions GGH, GH, and the common prior assumption all yield very similar conclusions. Hence I will consider all of these three under "condition GGH." Hence, as regards the assumptions of section 9, I will distinguish between when condition GGH holds and when K–L holds.

10.2. A guide to some of the literature. Below I provide an informal list of the results. More formal statements appear in section 12.

184

Case 1: Condition GGH and Assumption 8.4 hold

(a) *"Limit points of (type unconditional) beliefs about the future given the past are Nash equilibrium distributions." This result is proved in Nyarko (1993c).*

(b) *Suppose, in addition, that the common prior assumption 9.1 holds. Then the limit points of the "strategic representation of (type unconditional) beliefs are Nash equilibrium behavior-strategy profiles." This result is proved in Jordan (1991a).*

(c) *The sample path empirical distributions (averages of observed plays) and the beliefs of agents about the future conditional on the past "merge" over time. Hence, from part (a) the limit points of these empirical distributions are Nash equilibrium distributions. This result is proved in both Jordan (1992) and Nyarko (1993a). A few details are involved here, since in the result the empirical distributions must be constructed along subsequences for which the beliefs (not conditional on types) converge. It is this result that makes the conclusions of (a) and (b) interesting. Indeed, without this, one could genuinely ask: Who cares about the limit points of beliefs not conditional on types? The answer would be "because they are the same as the sample path averages of true play."*

Case 2: Condition K–L and Assumption 8.4 hold *Limit points of beliefs about the future conditional on types are Nash equilibrium distributions." This is proved by Kalai and Lehrer (1990). Since players know the strategies they choose, a player's belief about her own play conditional on her own type is equal to her actual play. Hence the Kalai–Lehrer result makes a stronger conclusion (about actual play) than the results of Case 1, which make conclusions about (type unconditional) beliefs about play. However, to obtain this stronger conclusion a stronger assumption is required (condition K–L as opposed to GGH).*

Case 3: Condition GGH holds but Assumption 8.4 fails *In this case all of the results of Case 1 above hold; however, wherever the word "Nash" appears in the statement of the results, one must insert in its place the word "correlated." In particular, when 8.4 fails, we obtain convergence as in Case 1 but to correlated equilibria and distributions rather than to Nash. The proof of this result for the zero-discount-factor case appears in Nyarko (1993a). (The extension to the positive-discount-factor case remains to be proved.)*

10.3. A guide to the examples of the next section.

10.3.1 Ex. 11.1 The example will obey condition GH but will violate K–L and will also violate the common prior assumption. The example will also obey

8.4. The emphasis of this example will be to illustrate that difference between convergence of beliefs (which occurs under condition GGH) and convergence of actual play (which requires the stronger condition K–L). Example 11.1 will illustrate the following:

  (i) *limit points of actual play or beliefs conditional on one's own types are not Nash equilibria; and*
 (ii) *limit points of beliefs about play not conditional on types are Nash equilibria.*

10.3.2 Ex. 11.2 *This example will obey condition GH but will violate K–L and will violate the common prior assumption. The example also violates 8.4. The emphasis of the example will be to show that a violation of 8.4 leads to correlated (as opposed to Nash) equilibria. The example will also emphasize the results Case 1(c) on empirical distributions. Specifically, the example illustrates the following:*

  (i) *limit points of actual play are not necessarily Nash equilibrium distributions;*
 (ii) *limit points of beliefs about play (whether conditional or not conditional on own types) are not necessarily Nash equilibrium distributions;*
(iii) *limit points of beliefs about play not conditional on own types are correlated equilibrium distributions;*
(iv) *limit points of beliefs about play conditional on own types are not necessarily correlated equilibrium distributions; and*
 (v) *the empirical distribution of play along each sample path and the beliefs about play not conditional on types "merge" over time.*

(Example 11.2 will have correlations in players' actions persisting over time. This example, which obeys condition GGH but fails K–L, is therefore a counterexample to Lehrer (1992) and shows that under the weaker condition GGH correlations may not "dissolve" over time.)

10.3.3 Ex. 11.3 The two-armed bandit problem in the context of a game *This example will obey conditions GH, GGH, K–L, and Assumption 8.4. The example will illustrate the following:*

  (i) *Despite the fact that I will conclude that limit points (of play or beliefs) are "Nash," we may still obtain the "two-armed bandit" effect of single-agent decision theory (see, e.g., Rothschild 1974). That is, one agent may persistently choose the "bad" action (or the inferior arm of a slot machine in the two-armed bandit story); this action results in no information being obtained as to the opponent's true type and hence this "bad" action is chosen forever. That this observation should be true in the single-agent case is not surprising. It is a little surprising in the multi-agent case situation because of the fact that there we are concluding that limit points of actions and beliefs are Nash equilibria.*

(ii) All the results in the literature conclude that it is the limit of the continuation of the Kuhn (1953) "strategic representations of beliefs" (or KSRB's) that are Nash equilibria. (See section 12 for formal definitions.) This example will show that limit points of the continuation of the true strategies are not Nash equilibria–even under condition K–L and the common prior assumption. (This observation has also been made in Jordan 1991b.)

11. THE EXAMPLES

Example 11.1 (Nyarko 1993c) *Suppose there are three players, A, B, and C, each with two actions, L and R. Player i could be any type τ_i in the interval $T_i = [\underline{\tau}_i, \overline{\tau}_i]$ where $0 < \underline{\tau}_i < \overline{\tau}_i < \infty$. The utility or payoff function is given by the following matrix box, where in each box the payoffs are those for players A, B, and C respectively.*

		If player C goes L and player B goes				If player C goes R and player B goes	
		L	R			L	R
Player A	L	$1, 1, -\tau_C$	$0, 0, 0$	A	L	$-1, 0, 0$	$-\tau_A, 0, 0$
	R	$0, 1, 0$	$0, 0, 1$		R	$0, -\tau_B, 0$	$0, 0, 0$

Note that the complete-information game with a given fixed and known vector of types (τ_A, τ_B, τ_C) does not have a Nash equilibrium in pure strategies. Also notice that if any player chooses the action R, then that player receives a payoff of zero regardless of the actions the other players choose.

For each $i \in I$, let π_i be any probability distribution over the type space T. π_i will be player i's prior probability over the type space. (In the notation of the previous sections π_i will be the marginal of μ_i on the type space T.) At the beginning of the initial period, player i realizes her type, τ_i, and is given no other information (and in particular is not told of the types of the other players, τ_{-i}). Player i's belief about the other players' types will be given by the conditional probability, $\pi_i(. \mid \tau_i)$. I will suppose that for each i, π_i is a product measure over $T_A \times T_B \times T_c$; (this is 8.4). This implies that $\pi_i(. \mid \tau_i)$ is independent of τ_i. I will suppose further that π_i admits a strictly positive and continuous Lebesgue density function over T. Each player has a zero discount factor so at each date seeks to maximize her expected payoffs within that period.

187

Consider player A. Suppose she assigns probability p_{AB} (resp. p_{AC}) to player B (resp. C) choosing action L. The expected return to A choosing action L is

$$(11.1.1) \qquad p_{AB}p_{AC} - p_{AB}(1 - p_{AC}) - \tau_A(1 - p_{AC})(1 - p_{AB}).$$

Since A receives a payoff of zero if A plays R, player A will choose action L if $\tau_A < \tau_A^*$ where

$$(11.1.2) \qquad \tau_A^* = [p_{AB}p_{AC} - p_{AB}(1 - p_{AC})]/(1 - p_{AC})(1 - p_{AB})$$

(and where, whenever the denominator in this expression is zero, we define $\tau_A^* = +\infty$ or $-\infty$ depending upon whether the numerator in the expression is positive or negative). Similarly, if p_{ij} is the probability assigned by player i to the event that player j chooses the action L, then player i chooses action L (resp. R) if $\tau_i < \tau_i^*$ (resp. $\tau_i > \tau_i^*$), and player i is indifferent between actions L and R if $\tau_i = \tau_i^*$, where τ_A^* is as in (11.1.2) above, and where

$$(11.1.3) \quad \tau_B^* = [p_{BA}p_{BC} + (1 - p_{BA})p_{BC}]/[(1 - p_{BA})(1 - p_{BC})] \text{ and}$$

$$(11.1.4) \quad \tau_C^* = (1 - p_{CA})(1 - p_{CB})/[p_{CA}p_{CB}].$$

The vector of numbers $\tau^* = (\tau_A^*, \tau_B^*, \tau_c^*)$ determines the behavior of each player. I will construct my example so that there is agreement as to how each player will behave as a function of that player's type. However there will be imperfect information on what the types of each player actually are, and there will in general be disagreement about the relative likelihoods of each type (e.g., B and C may have different beliefs about A's type). Hence, in the example, the vector of critical types $\tau^* = (\tau_A^*, \tau_B^*, \tau_C^*)$ will be known to each player. Player i's belief about the type space is given by π_i. If player i knows the critical type vector, τ^*, this will determine i's belief about the actions of other players, $\{p_{ij}\}_{j\neq i}$. In particular, we will have

$$(11.1.5) \qquad p_{AB} = \pi_A([\underline{\tau}_B, \tau_B^*]) \quad \text{and} \quad p_{AC} = \pi_A([\underline{\tau}_C, \tau_C^*]);$$

$$(11.1.6) \qquad p_{BA} = \pi_B([\underline{\tau}_A, \tau_A^*]) \quad \text{and} \quad p_{BC} = \pi_B([\underline{\tau}_C, \tau_C^*]);$$

$$(11.1.7) \qquad p_{CA} = \pi_C([\underline{\tau}_A, \tau_A^*]) \quad \text{and} \quad p_{CB} = \pi_C([\underline{\tau}_B, \tau_B^*]).$$

Given any tuple of beliefs about the type space $\{\pi_i\}_{i\in I}$, the critical vector of types, $(\tau_A^*, \tau_B^*, \tau_C^*)$, will be determined by the simultaneous solution of the equations (11.1.2)–(11.1.7). (The existence theorem of Milgrom and Weber (1985) may be used to formally show that such critical types exist.) Each player i will have a belief over the vector of initial types and date-1 actions, $T \times S$, induced by π_i and the critical types, τ^*. Denote this measure by μ_i^1. Then player i's behavior as described by i's critical type, τ_i^*, is a best response to i's belief, μ_i^1.

At the beginning of date 2, the players will observe some vector of actions s^1. This will indicate to each player that the vector of types is in some rectangle $T_2 \equiv \prod_{i \in I} [\underline{\tau}_{2i}, \bar{\tau}_{2i}]$. (For example, if the vector $s^1 = (L, R, L)$ is observed, then the vector of types must lie in the set $T_2 = [\underline{\tau}_A, \tau_A^*] \times [\tau_B^*, \bar{\tau}_B] \times [\underline{\tau}_C, \tau_C^*]$.) Each player's posterior distribution over the type space will then be the prior conditional on this information.

For date 2 in history s^1, we may mimic the construction for date 1 to obtain some critical types $(\tau_{2A}^*, \tau_{2B}^*, \tau_{2C}^*)$ such that under the following behavior at date 2 each player is best-responding to her belief: Player i chooses action L at date 2 if her type, τ_i, is less than or equal to τ_{2i}^* and chooses the action R otherwise. This process may be continued in each and every period to construct critical types at each date n in every history, $(\tau_{nA}^*, \tau_{nB}^*, \tau_{nC}^*)$, in a manner similar to that obtained for date 1. It should of course be noted that the values of the critical types at each date depend upon the past history. Each player i will have beliefs over the vector of initial types, T, and play of the game over the infinite time horizon, induced by π_i and the critical types, $\{(\tau_{nA}^*, \tau_{nB}^*, \tau_{nC}^*)\}_{n=1}^{\infty}$. Denote this[2] measure by μ_i. Then player i's behavior as described by i's critical types, $\{\tau_{in}^*\}_{n=1}^{\infty}$ is a best response to i's belief, μ_i. In particular, if we suppose that each player begins the game with a belief given by μ_i, then under μ_i each player's action at each date is a best response to the beliefs determined by μ_i. However, since we allow for $\pi_i \neq \pi_j$, players' beliefs about the game may differ.

Whatever is the true vector of types, notice that players are choosing a pure strategy (L or R) at each date. (The set of types where players are indifferent at any date is a countable set and has probability zero under π_i for each i.) Any limit point of the players' actions will therefore be some vector of pure strategies. However, for the true game with given vector of types $(\tau_A, \tau_B, \tau_C) \in T$, there does not exist a Nash equilibrium in pure strategies. Hence, along each sample path the limit points of actions chosen by the players do not constitute a Nash equilibrium for the true game with the given vector of types $\tau \in T$. This therefore verifies the claim made in Section 10.3's Example 11.1(i).

In this example we allow the priors π_A, π_B, and π_C to differ; hence, this is a model without common priors. In particular, two players, say A and B, may disagree about the probabilities that the type of the third player, player C, lies in the set $[\tau_C, \tau_C^*]$; players A and B may therefore disagree over the probability that player C will choose the action L at any given date. Let p_{ij}^n denote the probability assigned by player i conditional on the observed history, s^{n-1}, to the event that at date n player j chooses the action L. Since the priors of players are different, in general $p_{ij}^n \neq p_{kj}^n$ for $i \neq k$ for all $n < \infty$. It can be shown as a general result due to Blackwell and Dubins (1963) that the probabilities assigned by any two players to the event that the third player chooses L merge in the following sense: If along some subsequence the belief of one player (i.e., *the probability assigned to the third player choosing action L) converges to some limit point, then along the same subsequence the belief of the other player will converge to the same limit point. In particular,*

$$(11.1.8) \qquad |p_{AC}^n - p_{BC}^n| \to 0, |p_{AB}^n - p_{CB}^n| \to 0,$$
$$\text{and } |p_{BA}^n - p_{CA}^n| \to 0 \quad \text{as } n \to \infty.$$

Fix a true vector of types $(\tau_A, \tau_B, \tau_C) \in T$ and a sample path s^∞. Suppose along some subsequence of dates $p_{AC}^n \to p_C^\infty$, $p_{AB}^n \to p_B^\infty$, and $p_{BA}^n \to p_A^\infty$. Since beliefs of players' merge, we conclude that along the given sample path and subsequence of dates, $p_{BC}^n \to p_C^\infty$, $p_{CB}^n \to p_B^\infty$, and $p_{CA}^n \to p_A^\infty$. The results of Nyarko (1993c) show that $(p_A^\infty, p_B^\infty, p_C^\infty)$ is a (mixed-strategy) Nash equilibrium for the normal form game with true vector of types (τ_A, τ_B, τ_C). So, loosely speaking, "the (type unconditional) beliefs of the players converge to a Nash equilibrium." (This is the claim made in 10.3.1(ii).)

Example 11.2 (Nyarko 1993) *Consider the following 3-person game-payoff matrix (used by Aumann (1964) in a slightly different context):*

0, 1, 3	0, 0, 0
1, 1, 1	1, 0, 0

2, 2, 2	0, 0, 0
2, 2, 0	2, 2, 2

0, 1, 0	0, 0, 0
1, 1, 1	1, 0, 3

Player A chooses the row (Top or Bottom), B chooses the column (Left or Right), and C chooses the matrix (First, Middle, or Third). Using dominance arguments, it is easy to see that in any Nash equilibrium to the above normal form game, player A chooses Bottom, player B chooses Left column, and player C randomizes (with any probabilities) between the First and Third matrices.

Let ω be a realization from infinitely many independent and identical coin-tossing experiments where an outcome from {Heads, Tails} is chosen with equal probability. Hence ω is an element of {Heads, Tails}$^\infty$. Players A and B are told of the realization of ω. We may consider player A's "type" to be $\tau_A = \omega$, player B's "type" to be $\tau_B = \omega$, so that $\tau_A = \tau_B$. Player C is not told of ω but knows how it is chosen (i.e., C knows the distribution of ω). Player C has a trivial type space (consisting of a singleton element, say, representing "no information"). Let ω_n denote the n-th coordinate of ω. In particular, $\omega_n \in \{Heads, Tails\}$. Consider the following strategies for the players: At each date n, player A of type $\tau_A = \omega$ chooses Top at date n if $\omega_n = Heads$ and Bottom if $\omega_n = Tails$. Player B of type $\tau_B = \omega$ chooses the left column if $\omega_n = Heads$ and Right if $\omega_n = Tails$. Player C chooses the Middle matrix of all the time. It should be easy to see that if any player believes the others are choosing actions in the manner just described, then it is optimal for that player to choose actions in the manner described above for that player. In particular, each player is choosing a best response at each date to her belief about the other players.

Under this behavior, observe that Middle matrix is chosen at each date. This is not a Nash equilibrium action for the true game. Hence, I conclude thus: The actions or play of the players need not converge to a Nash equilibrium of the true game. This verifies the claim made in 10.3.2(i). Now consider the beliefs of players about the date n play, conditional on the history of the game from date 1 through date $n - 1$. These beliefs will assign probability one to Middle matrix being played. Middle is not part of any Nash equilibrium. In particular, we may make the following conclusion: The beliefs of players about the future play of the game conditional on the past (either conditional or not conditional on players' own realized types) need not converge to a Nash equilibrium of the true game. This verifies the claim of 10.3.2(ii).

This example appears to contradict the conclusions of Jordan (1991a and b) and Nyarko (1993c), where convergence of beliefs to a Nash equilibrium was proved. However, notice that the players' beliefs about the types of others are not independent of their own type; indeed, we have extreme dependence with $\tau_A = \tau_B$. Hence the independence assumptions used in the just-mentioned papers, and in particular Assumption 8.4, are violated.

However, notice that the behavior I have described in this example is actually a correlated equilibrium. Players A and B use the outcomes of the coin tosses to coordinate their actions across (Top, Left) and (Bottom, Right). The outcomes (Top, Left, Middle) and (Bottom, Right, Middle) with probability 1/2 each constitutes a correlated-equilibrium distribution. In particular, suppose a "principal" (or correlating device) "suggests" that player A should play action Top. Then A knows, in the correlated equilibrium, that B and C will play Left and Middle, respectively, so it is optimal for A to follow the suggestion of the principal. Similarly for B. In the correlated equilibrium, the only action that will be suggested to player C is the action Middle. C assigns equal probability to A and B choosing action pairs (Top, Left) and (Bottom, Right). Hence following the suggestion of the principal is optimal. This verifies that (Top, Left, Middle) and (Bottom, Right, Middle) with probability 1/2 each is indeed a correlated-equilibrium distribution.

Note further that this distribution is also the belief of each player about the next-period play of the game not conditional on that player's own type. In particular, if we asked each player to predict the outcome of the future of the game conditional on the history of the game but not conditional on their own realized type, then each player would predict the play to be (Top, Left, Middle) and (Bottom, Right, Middle) with probability 1/2 each. Hence, for this example, we may make the following conclusion: Limit points of beliefs of players about the future of the game conditional upon the past of the game but not conditional upon own types are correlated-equilibrium distributions for the true game. This illustrates the assertions of 10.3.2(iii).

Beliefs about the future of the game conditional upon the past and conditional upon own types do not converge to the set of correlated equilibria. Indeed, A's belief about the date-N play of the game conditional upon A's realized type and the history preceding date N is either that (Top, Left, Middle) will occur with probability one or that (Bottom, Right, Middle) will occur with probability one. (Which will occur is of course determined completely by the date-N coordinate, ω_N, of A's realized type ω.) However, neither of the outcomes (Top, Left, Middle) with probability one nor (Bottom, Right, Middle) with probability one is a correlated equilibrium, since in either case player C will be choosing a suboptimal action. In particular, I conclude that the beliefs of players about the future of the game conditional upon the past of the game and conditional upon own types need not converge to a correlated-equilibrium distribution for the true game. This verifies the claim made in 10.3.2(iv).

Let us now look at actual play again. We may invoke the strong law of large numbers to conclude that for almost every sample path, in each sufficiently long history or play of the game, the outcome (Top, Left, Middle) will occur for approximately as many periods as the outcome (Bottom, Right, Middle). In particular, the average number of times each outcome will occur will, in the limit, be equal to 1/2 for almost every sample path. I argued earlier that this outcome is the same as the limit point of beliefs of players (not conditional on own types) and, in particular, is a correlated equilibrium for the true game. Define the empirical distribution of play to be the distribution (or histogram) obtained by taking the average number of occurrences of each action in the past history. We may therefore conclude the following: The empirical distribution of play converges to a correlated equilibrium distribution for the true game. Further, the beliefs of players not conditioning on types and the empirical distribution of play converge to the same limit points over time. This illustrates the assertion of 10.3.2(v), and also illustrates the assertion of Case 1(c) for the case when the independence assumption 8.4 fails, as discussed in Case 3.

Example 11.3 (the bandit problem) *There are two players, A and B, each with two actions, $\{L, R\}$ and $\{l, r\}$ respectively. Player A has a utility function independent of actions. Player B's payoffs are given by $u_B(L, 1) = 1$, $u_B(R, r) = 8$, $u_B(R, l) = u_B(L, r) = 0$. Player B has a discount factor of $\delta = 1/2$. Player A may be one of two possible "types," α' and α'' respectively. Player B assigns probability of ϕ and $1 - \phi$ to A's type being α' or α'' respectively. I will take ϕ to be very close to one, and indeed in my computations below, any $\phi \in (8/9, 1)$ will do. I define the strategies $f_{\alpha'}$, $f_{\alpha''}$, f_β for player A type α', player A type α'', and player B as follows: $f_{\alpha'}$ plays L in each period regardless of the history. Strategy $f_{\alpha''}$ plays L in the first period (i.e., in the null history) and plays L in every history where player B has never played*

r; $f_{\alpha''}$ plays R in any history where at some previous date player B has played r. Strategy f_β plays l in each and every period.

I will first show that with the behavior strategies $(f_{\alpha'}, f_{\alpha''}, f_\beta)$ for the game where $\{\alpha', \alpha''\}$ has ex ante probabilities $\{\phi, (1 - \phi)\}$, each player of each type is best-responding. Player A is of course trivially choosing a best-response to f_β. If B uses f_β, then the outcome will be $(L, 1)$ in each period regardless of which of $f_{\alpha'}$ or $f_{\alpha''}$ player A is using. This results in a payoff to B of l in each period, so a discounted payoff of $1/(1 - \delta) = 2$. Suppose the B attempts an alternative strategy. In particular, suppose that at date l player B decides to play R. Since in the first period player A will play L regardless of her type, player B receives a payoff of 0 at date l. At the beginning of date 2 player B will still not know A's type. A's type will be revealed at the end of date 2 regardless of what B does at the beginning of date 2. The expected payoffs to B at date 2 are ϕ or $8(1 - \phi)$ depending upon whether B plays l or r respectively. Since $\phi > 8/9$, player B's optimal date-2 action is to choose l. At the beginning of date 3, player B will know precisely A's type. B will play l or r in each and every subsequent period depending upon whether A is revealed to be of type α' or α'' respectively; this will result in a sum of discounted payoffs to B (from date 3 onward) of $\delta^2/(1 - \delta) = 1/2$ or $8\delta^2/(1 - \delta) = 4$ respectively. Hence, if at date l B chooses the action r then B's optimal sum of expected discounted returns will be $0 + \delta[8(1 - \phi)] + \phi(1/2) + (1 - \phi)(4) = 8 - 7.5\phi$, which is less than 2 for all $\phi > 8/9$. Hence, the strategy of choosing r in any period to reveal the type of A is not optimal. Strategy f_β is therefore the unique best response for B.

It should be clear that this example obeys the independence assumption 8.4 and conditions GH, GGH, and K–L. It is also easy to check that $(f_{\alpha''}, f_\beta)$ is not a Nash equilibrium. Indeed, if B knows that A is of type α'', then when B uses f_β, B receives a discounted payoff of 2. On the other hand, when B plays l in the first period and plays r in each subsequent period, then B receives a payoff of $1 + [8\delta/(1 - \delta)] = 9$, which dominates the payoff from f_β. Further, on the event that A's true type is α'', each history of play, s^N, has $(L, 1)$ played at each date. For such histories the continuation strategies $f_{\alpha''}(. \mid s^N)$ and $f_\beta(. \mid s^N)$ are the same as $f_{\alpha''}$ and f_β respectively. Hence, the continuation strategies do not converge to a Nash equilibrium strategy profile.

This may appear at first sight a contradiction to the assertions made in sections Case 1 and Case 2. Those sections made conclusions about the (Kuhn 1953) strategic representation of beliefs (KSRB's) (these are formally defined in section 12); they do not make conclusions about the continuations of true or actual strategies. Indeed, it may be instructive to construct the KSRB's for this example. Player B's strategy is of course f_β with action l in each period. Player B's KSRB of player A's play, which we denote by f_A^*, is the following:

L is chosen in the first period and at each date when B has never previously chosen the action r. If at some date B chooses the action r for the first time, A flips a (biased) coin with probability of "heads" equal to ϕ and "tails" equal to $1 - \phi$. If the coin-toss realization is "heads" (resp. "tails"), A chooses the action L (resp. R) in each and every subsequent period. Now let us suppose that B is certain that A is using the strategy f_A^*. It is easy to check that the best response to this known strategy is f_β, which results in the choice of the action 1 in each and every period. Hence (f_A^*, f_β) is a Nash equilibrium for the repeated game. However in the real game played, with probability $(1 - \phi)$, player A plays $f_{\alpha''}$, which with f_β does not constitute a Nash equilibrium for the repeated game.

Further, note that this example is similar to the two-armed bandit problem of single-agent decision theory (see, e.g., Rothschild [1974]). In particular, if B were sure that A was of type α'', then B would choose action r in each period. However, because B assigns high probability to A's being of the first type, B never chooses action r. In so doing, B is never able to determine the true type of A. In the event that A's true type is α'', then an outside observer with knowledge of A's type will conclude that B's action r is the "wrong" action.

12. A FORMAL STATEMENT OF THE ASSERTIONS OF SECTION 10

Fix any collection of ex ante subjective beliefs of players, $\{\mu_i\}_{i \in I}$. For each date N, fix any version $\mu_i(. \mid s^N)$ of the conditional probability of μ_i given the history s^N. Let $\mu_{iN}(ds^{N++} \mid s^N)$ denote the probability distribution over the "future," $s^{N++} \equiv \{s_{N+1}, s_{N+1}, \ldots\} \in S^\infty$ conditional on the "past," s^N, induced by the measure $\mu_i(. \mid s^N)$. The norm $\| \cdot \|$ denotes the total variation norm on S^∞; that is, given $p, q \in \mathcal{P}(S^\infty)$,

$$(12.1) \qquad \|p\| \equiv \mathrm{Sup}_E |p(E) - q(E)|$$

where the supremum is over (Borel measurable) subsets E of S^∞. Define

$$(12.2) \quad W \equiv \{(\tau, f, s^\infty) \in T \times F \times S^\infty : \lim_{n \to \infty} \|\mu_{iN}(ds^{N++} \mid s^N)$$
$$- \mu_{jN}(ds^{N++} \mid s^N)\| = 0 \forall i, j \in I\}.$$

$$(12.3) \quad W^{**} \equiv \{(\tau, f, s^\infty) \in T \times F \times S^\infty : \lim_{n \to \infty} \|\mu_{iN}(ds^{N++} \mid s^N, \tau_i)$$
$$- \mu_{jN}(ds^{N++} \mid s^N, \tau_j)\| = 0 \forall i, j \in I\}.$$

The set W is the event in which the players' beliefs about the future, $s^{N++} \equiv \{s_{N+1}, s_{N+2}, \ldots\}$, given the past, s^N, (and not conditional on types) "merge" as the date N tends to infinity. In particular, on W the sequences $\{\mu_{iN}(ds^{N++} \mid s^N)\}_{n=1}^\infty$ and $\{\mu_{jN}(ds^{N++} \mid s^N)\}_{n=1}^\infty$ (which are sequences in $\mathcal{P}(S^\infty)$) have the same limiting behavior. If along a subsequence of dates one of the sequences converges, then the other will also converge along that subsequence, and the convergence will be to the same limit. The set W^{**} is the same as W except that in W^{**} we condition on players' own types.

Let $\{x^n\}_{n=1}^{\infty}$ be a sequence in some metric space X. Let D be any subset of X. We write $x^n \to^c D$ if every cluster point of $\{x^n\}_{n=1}^{\infty}$ lies in the set D. In the theorem below, $\mathcal{P}(S^{\infty})$ is endowed with its weak topology. F_{in} is endowed with its topology of pointwise convergence, and F_i and F are endowed with their respective product topologies. Define

$$(12.4) \qquad G \equiv \{(\tau, f, s^{\infty}) \in T \times F \times S^{\infty}$$
$$| \mu_{iN}(ds^{N++} \mid s^N) \to^c ND(\theta(\tau)) \forall i \in I\} \text{ and}$$

$$(12.5) \qquad G^{**} \equiv (\{(\tau, f, s^{\infty}) \in T \times F \times S^{\infty}$$
$$| \mu_{iN}(ds^{N++} \mid s^N, \tau_i) \to^c ND(\theta(\tau)) \forall i \in I\}.$$

The set G (resp. G^{**}) is the set in which limit points of each player's belief not conditional on own type (resp. conditional on own type) is a Nash equilibrium distribution.

For each $i \in I$, we define equivalence class relation, $\sim$, on F_i as follows: For each f_i and $f_i' \in F_i$, $f_i \sim f_i'$ if for all $f_{-i} \in F_{-i}, m(f_i, f_{-i}) = m(f_i', f_{-i})$ (where $m(f)$ is as in (3.1.1)). Let $F_i \sim$ denote the set of equivalence classes of $\sim$. From Kuhn's (1953) theorem (and Aumann [1964] for the infinite horizon case), we may conclude that there is a function $\kappa_i : \mathcal{P}(F_i) \to F_i \sim$ such that for any (mixed strategy) $\phi_i \in \mathcal{P}(F_i)$ and any $f_i \in \kappa(\phi_i)$ and any $f_{-i} \in F_{-i}$, the probability distribution on S^{∞} induced by ϕ_i and f_{-i} is equal to $m(f_i, f_{-i})$. The behavior strategy $f_i \in \kappa(\phi_i)$ is said to be *realization-equivalent* to, or a strategic representation of, the mixed strategy ϕ_i.

Fix any collection of ex ante subjective beliefs of players $\{\mu_i\}_{i \in I}$. Fix any $i, j \in I$. Recall that we defined $\mu_i(df_j)$ to be the marginal of μ_i on F_j. Fix any $f_j^{*i} \in F_j$ that is realization-equivalent to $\mu_i(df_j)$, that is,

$$(12.6) \qquad\qquad f_j^{*i} \in \kappa(\mu_i(df_j)).$$

Next define

$$(12.7) \qquad\qquad f^{*i} = \{f_j^{*i}\}_{j \in I}.$$

From 8.3 and 8.4, μ_i is a product measure on $\prod_{j \in I}[T_j \times F_j]$. Consider an outside observer with belief $\mu_i \in \mathcal{P}(T \times F \times S^{\infty})$ who never observes the types of the players. Such an outside observer will have a belief over the strategy space equal to the product measure $\mu_i(df) \equiv \prod_{j \in I} \mu_i(df_j)$. This is realization- equivalent to the strategy vector f^{*i} defined in (12.7). In particular, the marginal of μ_i on S^{∞}, $\mu_i(ds^{\infty})$, will equal the distribution $m(f^{*i})(ds^{\infty})$

over S^∞ induced by the behavior strategy f^{*i}. Hence, we may refer to f^{*i} as the *Kuhn strategic representation of beliefs* (or KSRB) of $\mu_i(df)$. I stress here that in this representation $\mu_i(df)$ does *not* condition on types. When the common prior assumption holds (i.e., when $\mu_i = \mu_j \forall i, j \in I$), the players have the same KSRB's (i.e., $f^{*i} = f^{*j} \forall i, j \in I$).

The shift operator $\sigma_{iN} : S^N \times F_i \to F_i$ is defined by setting for each $s^N \in S^N$ and $f_i \in F_i$, $\sigma_{iN}(s^N, f_i) \equiv f_i'$ where the date n coordinate of f_i' is defined by $f_{in}'(\cdot) \equiv f_{iN+n}(S^N, \cdot)$. If f^{*i} is the KSRB of the outside observer with belief μ_i who does not observe players' types, then $f_n^{*i} \equiv \sigma_{iN}(s^N, f^{*i})$ is the KSRB of that outside observer's belief about the future of the game following the date-N history s^N. Note well that f_N^{*i} depends upon the history s^N. If $m(\cdot)$ is as in (3.1.1), then $m(f_N^{*i})$ is the distribution over the future following s^N according to our outside observer's belief. Note that the set G in (12.4) is the set in which $m(f_N^{*i}) \to^c ND(\theta(\tau)) \forall i \in I$.

I now use the terminology we have just introduced to rewrite the corresponding assertions of section 10. I also formally state the "counterexamples" that have been established in section 11. Recall that μ^{**} denotes the true ex ante distribution (as in section 6) and μ^* is obtained from condition GGH.

Case 1(a)	Nyarko (1993c): $\mu^*(G \cap W) = 1$.
Case 1(b)	Jordan (1991b): $\mu^{**}(\{(\tau, f, s^\infty) \in T \times F \times S^\infty \| f_N^{*i}$ $= f_N^{*j} \to^c N(\theta(\tau)) \forall i, j \in I\}) = 1$.
Case 2	Kalai and Lehrer (1990): $\mu^{**}(G^{**} \cap W^{**}) = 1$.
10.3.1(i)	Ex. 11.1: $\mu^{**}(G^{**}) = 0$
10.3.2(ii)	Ex. 11.2: $\mu^{**}(G^{**}) = \mu^{**}(G) = 0$.
10.3.3(ii)	Ex. 11.3: $\mu^{**}(\{(\tau, f, s^\infty) \in T \times F \times S^\infty \|$ $\{\sigma_{iN}(s^N, f_i)\}_{i \in I} \to^c N(\theta(\tau))\}) = 0$.

NOTES

1. By appropriately defining the type space, this can be shown to be without loss of generality.
2. Notice that I have constructed μ_i to be a probability measure over $T \times S^\infty$ and not over $T \times F \times S^\infty$, as in the previous sections. However, since the discount factor is zero, this is really all we need. We may always extend this to a measure on $T \times F \times S^\infty$ by supposing that when the type vector is τ players choose the behavior-strategy profile f_τ, where f_τ is any behavior strategy that plays like $\mu_i(ds^\infty \mid \tau)$.

REFERENCES

Aumann, R. 1964. "Mixed and Behavior Strategies in Infinite Extensive Games." *Advances in Game Theory. Annals of Mathematical Studies* 5:627–50.

Blackwell, D., and L. Dubins. 1963. "Merging of Opinions with Increasing Information." *Annals of Mathematical Statistics* 38:882–86,

Blume, L., and D. Easley. 1984. "Rational Expectations Equilibrium: An Alternative Approach." *Journal of Economic Theory* **34**:116–29.

1992. "What Has the Rational Learning Literature Taught Us." Unpublished manuscript, Cornell University.

Harsanyi, J. C. 1967, 1968. "Games with Incomplete Information Played by Bayesian Players." Parts 1, 2, 3. *Management Science* Vol. 14, nos. 3, 5, 7., pp. 159–82, 320–34, 486–502.

Jordan, J. S. 1991a. "Bayesian Learning in Normal Form Games." *Games and Economic Behavior*, **3**:60–81.

1991b. "Bayesian Learning in Repeated Games." Unpublished manuscript, University of Minnesota.

1992. "Bayesian Learning in Games: A NonBayesian Perspective." Unpublished manuscript, University of Minnesota.

Kalai, E., and E. Lehrer. 1990. "Bayesian Learning and Nash Equilibrium." Unpublished manuscript, Northwestern University.

Kiefer, N., and Y. Nyarko. 1993. "Savage Bayesian Models of Economics." In *Essays in Learning and Rationality in Economics and Games*, eds. A. Kirman and M. Salmon. Basil Blackwell Press, forthcoming.

Kuhn, H. 1953. "Extensive Form Games and the Problem of Information," *Annals of Mathematics*, Study No. 193–216.

Lehrer, E. 1992. "Dissolving Correlations." Unpublished manuscript, Northwestern University.

Milgrom, P., and R. Weber. 1985. "Distributional Strategies for Games with Incomplete Information." *Mathematics of Operations Research* **10**(4):619–32.

Nyarko, Y. 1993a. "Bayesian Learning in Normal Form Games Leads to Correlated Equilibria." *Economic Theory* (forthcoming).

1993b. "The Savage–Bayesian Foundations of Economic Dynamics." Unpublished manuscript, New York University.

1993c. "Bayesian Learning without Common Priors and Convergence to Nash Equilibrium." Unpublished manuscript, New York University.

Rothschild, M. 1974. "A Two-Armed Bandit Theory of Market Pricing." *Journal of Economic Theory* **9**:185–202.

Savage, L. 1954. "The Foundations of Statistics." New York: Wiley.

10

Chaos and the explanatory significance of equilibrium: Strange attractors in evolutionary game dynamics

BRIAN SKYRMS

1. INTRODUCTION

The classical game theory of von Neumann and Morgenstern (1947) is built on the concept of equilibrium. I will begin this essay[1] with two more or less controversial philosophical claims regarding that equilibrium concept:

1: The explanatory significance of the equilibrium concept depends on the underlying dynamics.
2: When the underlying dynamics is taken seriously, it becomes apparent that equilibrium is not the central explanatory concept.

With regard to the first thesis, let me emphasize a point first made by von Neumann and Morgenstern themselves. Their theory is a static theory that discusses the nature and existence of equilibrium but does not address the question: "How is equilibrium reached?" The explanatory significance of the equilibrium concept, however, depends on the plausibility of the underlying dynamics that is supposed to bring players to equilibrium. One sort of story supposes that the decision makers involved reach equilibrium by an idealized reasoning process that requires a great deal of common knowledge, godlike calculational powers, and perhaps allegiance to the recommendations of a particular theory of strategic interaction. Another kind of story – deriving from evolutionary biology – views game-theoretic equilibria as fixed points of evolutionary adaptation, with none of the rational idealization of the first story. The power of game theory to explain a state of affairs as an equilibrium thus depends on the viability of a dynamical scenario appropriate to the situation in question, which shows how such an equilibrium would be reached.

It is well known that the problem is especially pressing in an area of game theory that von Neumann and Morgenstern did not emphasize: the theory of non–zero sum, noncooperative games. Here, unlike the zero-sum case, many nonequivalent equilibria are possible. If different decision makers aim

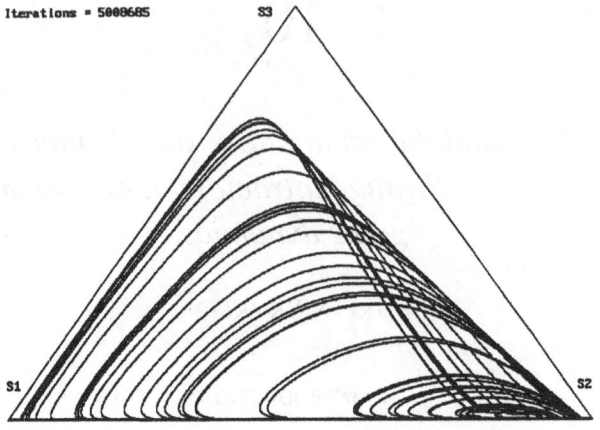

Figure 1. Parameter = 5

for different equilibria, then the joint result of their actions may not be an equilibrium at all. Thus the dynamics must bear the burden of accounting for *equilibrium selection* by the players, because without an account of equilibrium selection, the equilibrium concept itself loses its plausibility.

Once one has asked the first dynamical question "How is equilibrium reached?" it becomes impossible not to ask the more radical question: "Is equilibrium reached?" Perhaps it is not. If not, then it is important to canvass the ways in which it may not be reached and explore complex, nonconvergent behavior permitted by the underlying dynamics. This essay will take a small step in that direction.

In particular, I will present numerical evidence for extremely complicated behavior in the evolutionary game dynamics introduced by Taylor and Jonker (1978). This dynamics, which is based on the process of replication, is found at various levels of chemical and biological organization (Hofbauer and Sigmund 1988). For a taste of what is possible in this dynamics with only four strategies, see the "strange attractor" in Figure 1. This is a projection of a single orbit for a four-strategy evolutionary game onto the three simplex of the probabilities of the first three strategies. A strange attractor cannot occur in the Taylor–Jonker flow in three strategy evolutionary games, because the dynamics takes place on a two-dimensional simplex. Zeeman (1980) leaves it open as to whether strange attractors are possible in higher dimensions or not. This chapter presents strong numerical evidence for the existence of strange attractors in the lowest dimension in which they could possibly occur.

The plan of the paper is as follows: Sections 2, 3, and 4 introduce key concepts of games, dynamics, and evolutionary-game dynamics. Section 5 describes a four-strategy evolutionary game that gives rise to chaotic dynamics, and the bifurcations that lead to chaos as the parameters of the model are varied. Section 6 gives a stability analysis of the equilibria encountered along the road to chaos described in Section 5. Section 7 describes the numerical calculation of Liapunov exponents. Section 8 indicates some related literature and discusses the relation to Lotka–Volterra ecological models. My second philosophical claim will be discussed in Section 9.

2. GAMES

We will be concerned with finite, noncooperative, normal form games. There are a finite number of players and each player has a finite number of possible strategies. Each player has only one choice to make and makes it without being informed of the choices of any other players. The games are to be thought of as noncooperative. There is no communication or precommitment before the players make their choices. Each possible combination of strategies determines the payoffs for each of the players.

A specification of the number of players, the number of strategies for each player, and the payoff function determines the game. A *Nash equilibrium* of the game is a strategy combination such that no player does better on any unilateral deviation. We extend players' possible acts to include randomized choices at specified probabilities over the originally available acts. The new randomized acts are called mixed strategies, and the original acts are called pure strategies. The payoffs for mixed strategies are defined as their expected values using the probabilities in the mixed acts to define the expectation (and assuming independence between different players' acts). We will assume that mixed acts are always available. Then every finite, noncooperative, normal form game has a Nash equilibrium.

The game in Example 1 has two Nash equilibria in pure strategies, one at ⟨bottom, right⟩ and one at ⟨top, left⟩. Intuitively, the former equilibrium is, in some sense, highly unstable, and the latter equilibrium is the only sensible one.

Example 1

1, 1	0, 0
0, 0	0, 0

Selten (1975) introduced the notion of a *perfect equilibrium* to capture this intuition. He considers *perturbed* games wherein each player, rather than simply choosing a strategy, chooses to instruct a not perfectly reliable agent as to which strategy to choose. The agent has some small nonzero probabilities for mistakenly choosing each of the strategies alternative to the one he was instructed to choose. Probabilities of mistakes of agents for different players are uncorrelated. An equilibrium in the original game, which is the limit of some sequence of equilibria in perturbed games as the probability of mistakes goes to zero, is called a (trembling-hand) perfect equilibrium. In any perturbed game for Example 1, there is only one equilibrium, with row and column instructing their agents to play top and left and their agents doing so with probability of one minus the small probability of a mistake.

Classical game theory is intended as a theory of strategic interaction between rational, human payoff maximizers. It has sometimes been criticized as incorporating an unrealistically idealized model of human rationality. Maynard Smith and Price (1973) found a way to apply game theory to model conflicts between animals of the same species. The rationale obviously cannot be that snakes or mule deer are hyperrational, but rather that evolution is a process with a tendency in the direction of increased payoff where payoff is reckoned in terms of evolutionary fitness. A rest point of such a process must be an optimal point. The insight that just such a tendency may be enough to make rational choice theory and game theory relevant can be carried back to the human realm and accounts for much of the current interest in dynamic models of learning and deliberation in game-theoretic contexts.

Maynard Smith and Price are interested in providing an evolutionary explanation of "limited-war" conflicts between members of the same species, without recourse to group selection. The key notion they introduce is that of a strategy that would be a stable equilibrium under natural selection, an *evolutionarily stable strategy*. If all members of the population adopt that strategy, then no mutant can invade. Suppose that there is a large population, that contests are pairwise, and that pairing is random. Then the relevant payoff is the average change in evolutionary fitness of an individual, and it is determined by its strategy and the strategy against which it is paired. These numbers can be conveniently presented in a *fitness matrix* and can be thought of as defining the evolutionary game. The fitness matrix is read as giving row's payoff when playing against column.

Example 2

	R	H
R	2	-3
H	-1	-2

Thus in Example 2, the payoff to R when playing against R is 2 but when playing against H is -3. The payoff to H when playing against R is -1 and when

202

playing against H is -2. Here R is an evolutionarily stable strategy, because in a population in which all members adopt that strategy, a mutant who played H would do worse against members of the population than they would. Likewise, H is an evolutionarily stable strategy, since H does better against H than R does. Suppose, however, that a mutant could do exactly as well against an established strategy as that strategy against itself, but the mutant would do worse against itself than the established strategy. Then the established strategy should still be counted as evolutionarily stable, as it has greater average payoff than the mutant, in a population consisting of players playing it together with a few playing the mutant strategy. This is the formal definition adopted by Maynard Smith. Let $U(x|y)$ be the payoff to strategy-x player against strategy y. Strategy x *is evolutionarily stable* just in case $U(x|x) > U(y|x)$ or $U(x|x) = U(y|x)$ and $U(x|y) > U(y|y)$ for all y different from x. Equivalently, x is evolutionarily stable if:

1: $U(x|x) \geq U(y|x)$;
2: if $U(x|x) = U(y|x)$, then $U(x|y) > U(y|y)$.

The fitness matrix determines a symmetric payoff matrix for a two-person game – the symmetry deriving from the fact that only the strategies matter, not whether they are played by row or column – as shown in Example 3.

Example 3

2, 2	$-3, -1$
$-1, -3,$	$-2, -2$

An evolutionarily stable strategy is, by condition 1 above, a symmetric Nash equilibrium of the two-person, noncooperative game. Condition 2 adds a kind of stability requirement.

The formal definition of "evolutionarily stable strategy" applies to mixed strategies as well as pure ones, and some fitness matrices will have the consequence that the only evolutionarily stable strategy is a mixed one. This is illustrated in Example 4.

Example 4

	H	D
H	-2	2
D	0	1

Neither H nor D is an evolutionarily stable strategy, but a mixed strategy, M, of (1/3) H, (2/3) D is. This illustrates condition 2 in the definition of evolutionarily stable strategy. $U(x|M) = 2/3$ if x is H or D or any mixture of H and D. But an invader who plays H or D or a different mixture of H and D will do worse against herself than M does against her. For example, consider H as an invader. $U(H|H) = -2$ while $U(M|H) = -2/3$. The interpretation of mixed strategies as strategies adopted by each member of the population is the only one that makes sense of the characterization: if all members of the population adopt that strategy, then no mutant can invade. There is an alternative interpretation of mixed strategies in terms of proportions of a polymorphic population, all of whose members play pure strategies. The formal definition of evolutionarily stable strategy in terms of conditions 1 and 2 still makes sense on this reinterpretation of mixed strategies.

If we consider the two-person, noncooperative normal form game associated with a fitness matrix, an evolutionarily stable strategy, x, induces a symmetric Nash equilibrium $\langle x, x \rangle$ of the game that has certain stability properties. Earlier, we considered Selten's concept of perfect equilibrium, which rules out certain instabilities. Evolutionary stability is a stronger requirement than perfection. If x is an evolutionarily stable strategy, then $\langle x, x \rangle$ is a perfect, symmetric Nash equilibrium of the associated game, but the converse does not hold. In the game associated with the fitness matrix in Example 5, $\langle S2, S2 \rangle$ is a perfect equilibrium.[2]

Example 5

	$S1$	$S2$	$S3$
$S1$	1	0	-9
$S2$	0	0	-4
$S3$	-9	-4	-4

$S2$, however, is not an evolutionarily stable strategy, because $U(S1|S2) = U(S2|S2)$ and $U(S1|S1) > U(S2|S1)$.

The concepts of equilibrium and stability in game theory are quasi-dynamical notions. How do they relate to their full dynamical counterparts when game theory is embedded in a dynamical theory of equilibration?

3. DYNAMICS

The state of a system is characterized by a state vector, x, that specifies the values of relevant variables. (In the case of prime interest here, the relevant variables will be the probabilities of strategies in a game.) The dynamics of the system specifies how the state vector evolves in time. The path that a state vector describes in state space as it evolves according to the dynamics is called a trajectory, or orbit. Time can be modeled either as discrete or as continuous. For the former case, a deterministic dynamics consists of a map that may be specified by a system of difference equations:

$$x(t+1) = f(x(t)).$$

In the latter case, a deterministic dynamics is a flow that may be specified by a system of differential equations:

$$dx/dt = f(x(t)).$$

An *equilibrium point* is a fixed point of the dynamics. In the case of discrete time, it is a point, x, of the state space such that $f(x) = x$. For continuous time, it is a state, $x = \langle x_1, \ldots, x_i, \ldots \rangle$ such that $dx_i/dt = 0$, for all i. An equilibrium x is *stable* if points near to it remain near to it. More precisely, x is stable if for every neighborhood, V of x, there is a neighborhood, V' of x, such that if the state y is in V' at time $t = 0$, it remains in V for all time $t > 0$. An equilibrium x is *strongly stable* (or asymptotically stable) if nearby points tend toward it. That is, to the definition of stability we add the clause that the limit as t goes to infinity of $y(t) = x$.

An *invariant set* is a set, S, of points of the state space such that if the system starts at a point in S, then at any subsequent time the state of the system is still in S. A unit set is an invariant set just in case its member is an equilibrium. A closed, invariant set, S, is an *attracting set* if nearby points tend toward it, that is, if there is a neighborhood, V, of S such that the orbit of any point in V remains in V and converges to S. An *attractor* is an indecomposable attracting set. (Sometimes other conditions are added to the definition).

A dynamical system displays *sensitive dependence on initial conditions* at a point if the distance between the orbits of that point and one infinitesimally close to it increases exponentially with time. This sensitivity can be quantified by the *Liapunov exponent* of an orbit. For a one-dimensional map, $x(t+1) = f(x(t))$, this is defined as follows:[3]

$$\lambda = \lim_{n \to \infty} \frac{1}{n} \sum_{i=0}^{n-1} \log_2 \left| \frac{df}{dx} \text{ at } x_i \right|.$$

A positive Liapunov exponent may be taken as the mark of a *chaotic* orbit. For example, consider the following "tent" map:

$$x(t+1) = 1 - 2\left| \frac{1}{2} - x(t) \right|.$$

205

The derivative is defined, and its absolute value is 2 at all points except $x = 1/2$. Thus, for almost all orbits, the Liapunov exponent is equal to one.

An attractor for which the orbit of almost every point is chaotic is a *strange attractor*. For most known strange attractors – like the Lorenz attractor and the Rössler attractor – there is no mathematical proof that they are strange attractors, although the computer experiments strongly suggest that they are. The strange attractor in game dynamics that appears in Figure 1, and which will be discussed in Sections 5–7, has the same status.

4. GAME DYNAMICS

A number of different dynamical models of equilibration processes have been studied in economics and biology. Perhaps the oldest is the dynamics considered by Cournot (1897) in his studies of oligopoly. There is a series of production quantity settings by the oligopolists, at each time period of which each oligopolist makes her optimal decision on the assumption that the others will do what they did in the last round. The dynamics of the system of oligopolists is thus defined by a *best-response map*. A Nash equilibrium is a fixed point of this map. It may be dynamically stable or unstable, depending on the parameters of the Cournot model.

A somewhat more conservative adaptive strategy has been suggested by evolutionary game theory. Here we will suppose that there is a large population, all of whose members play pure strategies. The interpretation of a mixed strategy is now as a polymorphism of the population. Asexual reproduction is assumed for simplicity. We assume that individuals are paired at random, that each individual engages in one contest (per unit time), and that the payoff in terms of expected number of offspring to an individual playing strategy S_i against strategy S_j is U_{ij} – given in the ith row and jth column of the fitness matrix, U. The proportion of the population playing strategy S_j will be denoted by $pr(S_j)$. The expected payoff to strategy i is:

$$U(S_i) = \sum_j pr(S_j)U_{ij}.$$

The average fitness of the population is:

$$U(\text{Status Quo}) = \sum_i pr(S_i)U(s_i).$$

The interpretation of payoff in terms of Darwinian fitness then gives us a map for the dynamics of evolutionary games in discrete time:

$$pr'(S_i) = pr(S_i)\frac{U(S_i)}{U(\text{Status Quo})}$$

206

(where pr' is the proportion in the next time period). The corresponding flow is given by:

$$\frac{dpr(S_i)}{dt} = pr(S_i)\frac{U(S_i) - U(\text{Status Quo})}{U(\text{Status Quo})}.$$

As long as we are concerned, as we are here, only with symmetric evolutionary games, the same orbits are given by a simpler differential equation:

$$\frac{dpr(S_i)}{dt} = pr(S_i)[U(S_i) - U(\text{Status Quo})].$$

This equation was introduced by Taylor and Jonker (1978) to provide a dynamical foundation for the quasi-dynamical notion of evolutionarily stable strategy of Maynard Smith and Price (1973). It has subsequently been studied by Zeeman (1980), Hofbauer (1981), Bomze (1986), van Damme (1987), Hofbauer and Sigmund (1988), Samuelson (1988), Crawford (1989), reprinted in this volume and Nachbar (1990). It will be the dynamics considered in the example in the next section. It is worth noting that even though the Taylor–Jonker dynamics is motivated by a context in which the payoffs are measured on an absolute scale of evolutionary fitness, nevertheless the orbits in phase space (although not the velocity along these orbits) are invariant under a linear transformation of the payoffs. Thus the Taylor–Jonker dynamics may be of some interest in contexts for which it was not intended, where the payoffs are given in von Neumann–Morgenstern utilities.

Relying on the foregoing studies, I will briefly summarize some of the known relations between quasi-dynamical equilibrium concepts and dynamical equilibrium concepts for this dynamics. If $[M, M]$ is a Nash equilibrium of the two-person, noncooperative game associated with an evolutionary game, then M is a dynamic equilibrium of the Taylor–Jonker flow. The converse is not true, since every pure strategy is an equilibrium of the flow. However, if an orbit starts at a completely mixed point and converges to a pure strategy, then that strategy is a Nash equilibrium. Furthermore, if M is a stable dynamic equilibrium in the Taylor–Jonker flow, then $[M, M]$ must be a Nash equilibrium of the associated game. However if M is dynamically stable, $[M, M]$ need not be perfect, and if $[M, M]$ is perfect, then M need not be dynamically stable. If M is dynamically strongly stable (asymptotically stable), then $[M, M]$ must be perfect, but the converse does not hold. If M is an evolutionarily stable strategy in the sense of Maynard Smith and Price, then it is perfect, but the converse does not hold. We do have equivalence between evolutionarily stable strategy and strongly dynamically stable strategy in the special case of two-strategy evolutionary games, but already in the case of three strategies there can be a strongly dynamically stable polymorphic population that is not a mixed evolutionarily stable strategy. Thus, although there are important relations here between the quasi-dynamical and dynamical equilibrium concepts, they tend to draw the line at somewhat different places.

As an example of a third kind of dynamics, we mention the fictitious play of Brown (1951). Like the Cournot dynamics, there is a process in discrete time, at each stage of which each player plays a strategy that maximizes expected utility, according to her beliefs. But these beliefs are not quite so naive as those of the Cournot player. Rather than proceeding on the assumption that all other players will do just what they did last time, Brown's players form their probabilities of another player's next act according to the proportion of times that player has played that strategy in the past.[4] Brown interpreted his model as fictitious play, and Cournot interpreted his as real play, but either could just as well be interpreted the other way. Thorlund–Peterson (1990) studies a dynamics closely related to Brown's in the context of a Cournot oligopoly, where it is shown to have convergence properties superior to those of the Cournot dynamics. Brown's dynamics is driven by a simple inductive rule: Use the observed relative frequency as your probability. The basic scheme could be implemented using modified inductive rules. A class of simple Bayesian inductive rules that share the same asymptotic properties as Brown's rule is investigated in Skyrms (1991). For these models, if the dynamics converges, it converges to a Nash equilibrium in undominated strategies. For two-person games, such an equilibrium must be *perfect*. This contrasts with the Taylor–Jonker dynamics, in which an orbit can converge to a dynamically stable equilibrium, M, where $[M, M]$ is an imperfect equilibrium of the corresponding two-person noncooperative game. On the other hand, Shapley (1964) showed by example that fictitious play need not converge. Shapley's example is a 3-×-3, two-person game in which the dynamical attractor is a limit cycle. Recently Krishna and Sjöström (1995) have shown that cyclic nonconvergence (as illustrated by the Shapley example) is generic for a certain class of nonzero sum games.

5. THE ROAD TO CHAOS

In this section, I will focus on the Taylor–Jonker flow. Flows are usually better behaved than the corresponding maps, but we will see that this dynamics is capable of quite complicated behavior. Taylor and Jonker already note the possibility of nonconvergence because of oscillations in three-strategy evolutionary games. They consider the game whose fitness matrix, U, is given in Example 6 (where a is a parameter to be varied):

Example 6

	$S1$	$S2$	$S3$
$S1$	2	1	5
$S2$	5	a	0
$S3$	1	4	3

For $a = 1$, the completely mixed equilibrium serves as an example of an equilibrium that is dynamically strongly stable but is not an evolutionarily stable strategy. For $a < 3$, the equilibrium is strongly stable, but at $a = 3$ a qualitative change takes place. Now the mixed equilibrium is stable but not strongly stable. It is surrounded by closed orbits. At $a > 3$ the mixed equilibrium is unstable and the trajectories spiral outward to the boundary of the space. The change that takes place at $a = 3$ is a *degenerate* Hopf bifurcation. (See Guckenheimer and Holmes 1986, pp. 73 and 150 ff.) It is degenerate because the situation at $a = 3$ is not structurally stable. Any small perturbation of the value of a destroys the closed orbits. This is just about as wild as the dynamical behavior can get with three strategies. In particular, *generic* Hopf bifurcations are impossible here. (See Zeeman 1980 and Hofbauer 1981. Zeeman proves that a generic Hopf bifurcation is impossible for 3-strategy games and describes the structurally stable flows for such games under the assumption that is discharged in Hofbauer.) And chaotic strange attractors are not possible, because the flow takes place on a two-dimensional simplex.

However, with four strategies, we get the strange attractor pictured in Figure 1. (This is a projection of the three-dimensional simplex of probabilities for four strategies onto the two-dimensional simplex of the first three strategies. The three-dimensional structure, however, is fairly easy to see in the figure.) There is a route to this strange attractor that leads through a generic Hopf bifurcation. Consider the fitness matrix, U of Example 7 (where a is the parameter to be varied):

Example 7

-1	-1	-10	$1,000$
-1.5	-1	-1	$1,000$
a	.5	0	$-1,000$
0	0	0	0

Figures 1 through 6 are snapshots taken along the path to chaos as this parameter is varied. At $a = 2.4$ there is convergence to a mixed equilibrium as shown in

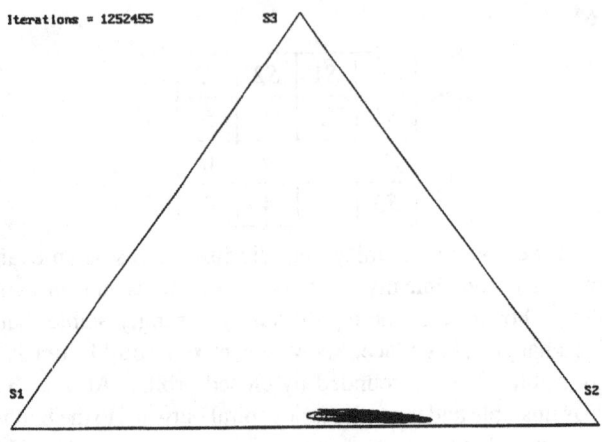

Figure 2. Parameter = 2.4

Figure 2. The orbit spirals in toward the mixed equilibrium that is visible as the white dot in the center of the orbit. As the value of a is raised, there is a generic Hopf bifurcation giving rise to a limit cycle around the mixed equilibrium. This closed orbit is structurally stable; it persists for small variations in the parameter. It is also an attracting set. This closed orbit is shown for $a = 2.55$ in Figure 3. As the value of the parameter is raised further, the limit cycle expands and then undergoes a period-doubling bifurcation. Figure 4 shows the cycle of period 2 at $a = 3.885$. This is followed by another period-doubling bifurcation, leading to a cycle of period 4 at $a = 4.0$, as shown in Figure 5. There are very long transients before the orbit settles down to this cycle. At $a = 5$, we get a transition to chaotic dynamics on the strange attractor shown in Figure 1. Raising the parameter to $a = 6$ leads to further geometrical complications in the strange attractor as shown in Figure 6.

Differential equations were numerically integrated in double precision using fourth-order Runge-Kutta method (Press et al. 1989). For Figures 1 through 4 and 6, a fixed step size of .001 was used. For Figure 5, a fixed step size of .01 was used. This was done on an IBM model 70 personal computer with a 387 math coprocessor. The projection of the orbit on the simplex of probabilities of the first three strategies was plotted to the screen in VGA graphics mode. For Figures 1 through 4, the first 50,000 steps ($= 50$ time units) were not plotted to eliminate transients. For Figure 5, the first 100,000 ($= 1,000$ time units) steps were omitted to eliminate very long transients. For Figure 6, only the first 1,000 steps were omitted. In each case, the total number of steps run is shown in the top left corner of the illustration. The screen was captured using the WordPerfect 5.1 "Grab" utility and printed on a Hewlett Packard LaserJet II.

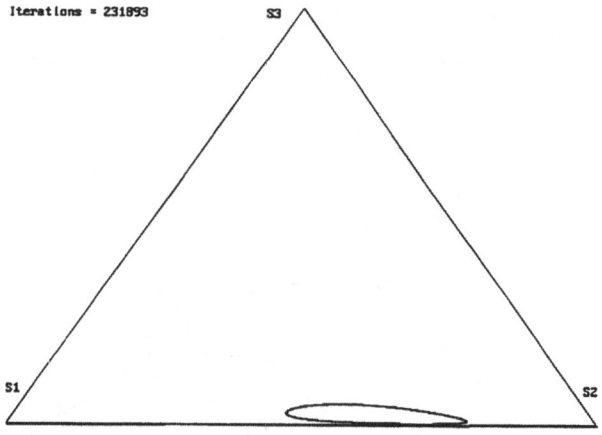

Figure 3. Parameter = 2.55.

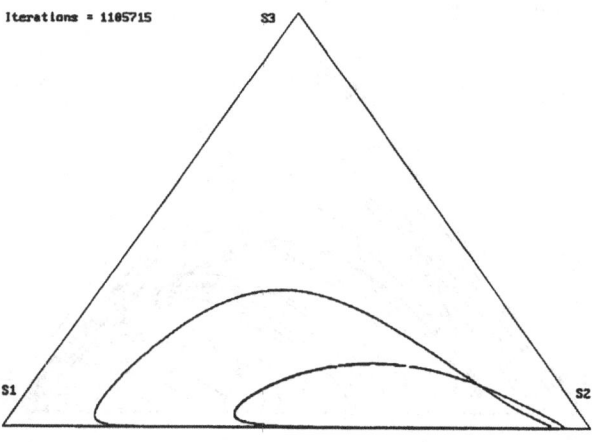

Figure 4. Parameter = 3.885.

6. STABILITY ANALYSIS OF EQUILIBRIA

As a supplement and a check on the graphical information presented in the previous section, the interior equilibrium points along the route to chaos were calculated in high precision (40 decimal places) using *Mathematica*. The Jacobian matrix of partial derivatives was then evaluated at the equilibrium point and its eigenvalues found. These are used in stability analysis of the equilibria. (See Hirsch and Smale 1974, ch. 6.) One of these eigenvalues will always

211

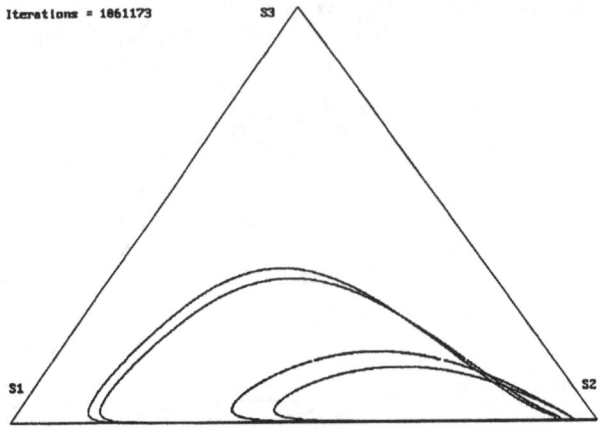

Figure 5. Paramter = 4.

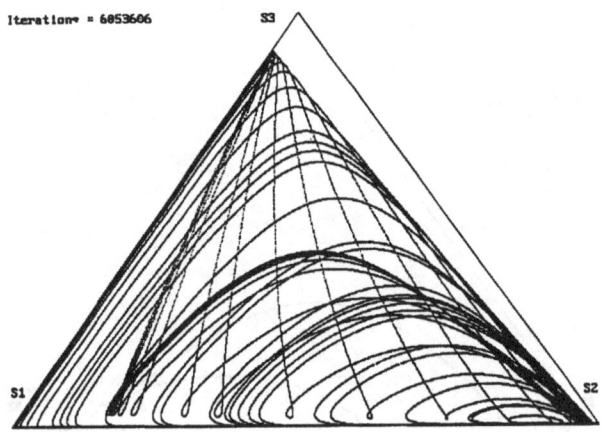

Figure 6. Parameter = 6.

be zero; it is an artifact of the constraint that probabilities add to one and is irrelevant to the stability analysis.[5]

For example, at $a = 2$ there is an interior equilibrium at:

$$x1 = 0.51363999543431115169501198493322680059
4$$
$$x2 = 0.45656888483049880150667731994064604497
2$$
$$x3 = 0.02853555530190617509416733249629037781
08$$
$$x4 = 0.00125556443328387170414336262983677662
367$$

212

and at this point the eigenvalues of the Jacobian matrix are found numerically to be:

$$- 0.8576108025804070626366657159513993308,$$
$$- 0.0562990388422014452825117944612367686$$
$$+ 0.2875123366474160989192752729129540\,4\,I$$
$$- 0.0562990388422014452825117944612367686$$
$$- 0.2875123366474160989192752729129540\,4\,I$$
$$- 5.4204572416321964652348917801112836 * 10^{-42}$$

The last eigenvalue is the insignificant zero eigenvalue. The significant eigenvalues all have negative real parts, indicating a strongly stable equilibrium, which attracts much in the way illustrated in Figure 2. Indeed at $a = 2.4$ – the situation actually illustrated in Figure 2 – the situation is qualitatively much the same. The equilibrium has moved to about:

$$x1 = 0.363942$$
$$x2 = 0.614658$$
$$x3 = 0.020219$$
$$x4 = 0.001181$$

(henceforth I suppress the full precision in reporting the results). The non-significant zero eigenvalue of the Jacobian matrix is numerically calculated at the order of 10^{-39}. The significant eigenvalues are approximately:

$$-0.9752593,$$
$$-0.001670447 + 0.26020784\,I$$
$$-0.001670447 - 0.26020784\,I$$

However, when we move to the limit cycle illustrated in Figure 3 at $a = 2.55$, the situation changes drastically. The equilibrium is now at approximately:

$$x1 = 0.328467$$
$$x2 = 0.653285$$
$$x3 = 0.018248$$
$$x4 = 0.001164$$

and the significant eigenvalues of the Jacobian matrix are:

$$-0.993192,$$
$$0.00572715 + 0.250703\,I,$$
$$0.00572715 - 0.250703\,I$$

The real eigenvalue is negative, but the imaginary eigenvalues have positive real parts. Thus the equilibrium is an unstable saddle, with the imaginary eigenvalues indicating the outward spiral leading to the limit cycle. A little trial and error in this sort of computation indicates that the *Hopf bifurcation*, where the real parts of the imaginary eigenvalues pass from negative to positive, takes place between $a = 2.41$ and $a = 2.42$, where the real parts of the imaginary eigenvalues are respectively about -0.001 and $+0.001$.

In the chaotic situation where $a = 5$ shown in Figure 1, the equilibrium has now moved to approximately:

$$x1 = 0.12574$$
$$x2 = 0.866212$$
$$x3 = 0.006956$$
$$x4 = 0.001070$$

The eigenvalues of the Jacobian matrix are:

$$-1.0267,$$
$$0.173705 + 0.166908\,I$$
$$0.173705 + 0.166908\,I$$

This still indicates a saddle-point equilibrium, but here, as shown in Figure 1, the orbit passes very close to this unstable equilibrium point.

7. NUMERICAL CALCULATION OF LIAPUNOV EXPONENTS

Liapunov exponents were calculated numerically using the algorithm presented in Wolf et al. (1985, appendix A). This integrates the differential equations of the dynamical system to obtain a fiducial trajectory, and simultaneously integrates four copies of the linearized differential equations of the system with coefficients determined by the location on the fiducial trajectory, to calculate the Liapunov spectrum. The latter are started at points representing a set of orthonormal vectors in the tangent space and are periodically reorthonormalized during the process. In the calculation, logarithms are taken to the base 2. The code was implemented for the replicator dynamics by Linda Palmer. Differential equations were integrated in double precision using the IMSL Library integrator DIVPRK. The program was tested running it at $a = 2$, starting it on the attracting equilibrium. In this case, the spectrum of Liapunov exponents (when converted to natural logarithms) should just consist of the real parts of the eigenvalues of the Jacobian matrix evaluated at the equilibrium, which were discussed in the last section. The experimental results of a run from $t = 0$ to $t = 110,000$ were in agreement with the theoretical results up to four or five decimal places:

Experimental Results	Theoretical Results
-0.85761	-0.85761
-0.0563	-0.0563
-0.0563	-0.0563
$-3.4 * 10^{-6}$	0

The three negative exponents indicate the attracting nature of the equilibrium point, and the zero exponent corresponds to the spurious eigenvalue as explained in the last section.

For a limit cycle in three dimensions, the Liapunov spectrum should have the qualitative character $\langle 0, -, - \rangle$. The experimental results on the limit cycles at $a = 2.55, a = 3.885$, and $a = 4$ have the appropriate qualitative character. Dropping one spurious zero exponent, we are left with:

	$a = 2.55$	$a = 3.885$	$a = 4$
$L1$	0.000	0.000	0.000
$L2$	-0.020	-0.008	-0.004
$L3$	-1.395	-1.419	-1.423

For a strange attractor in three dimensions, the Liapunov exponents should have the qualitative character $\langle +, 0, - \rangle$. At $a = 5$, where visually we see the onset of chaos in Figure 1, the Liapunov spectrum was calculated on a number of runs on a number of computers, varying the reorthonormalization frequency and various parameters of the differential equation integrator. Dropping one spurious zero exponent, the following results are very robust:

$$L1 : 0.010$$
$$L2 : 0.000$$
$$L3 : -1.44$$

For a "gold-standard run" the equations were integrated from $t = 0$ to $t = 1,000,000$ with an error tolerance of 10^{-11}. On this run the zeros (both $L2$ and the spurious exponent) are zeros to six decimal places. Details of the convergence are shown graphically in Figures 7–10 (where one unit on the x axis represents 10,000 units of time). The positive value of the largest Liapunov exponent, $L1$, indicates that there has indeed been a transition to chaos.[6]

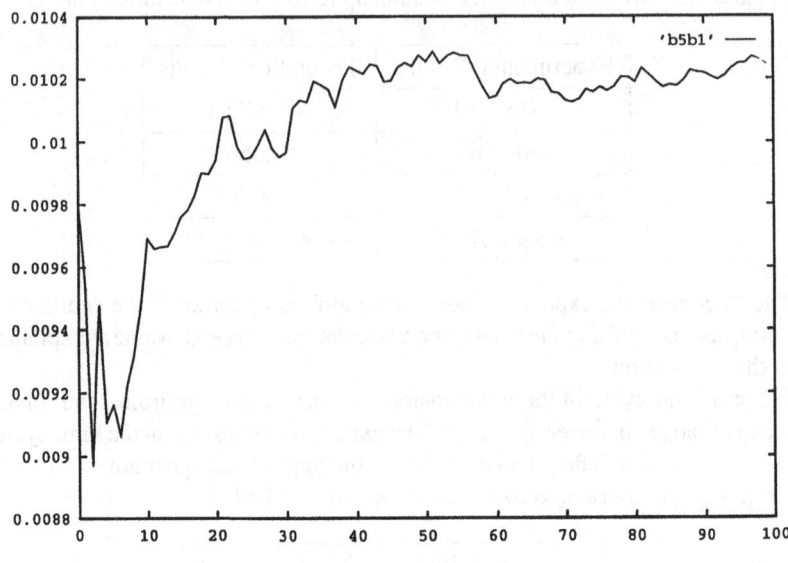

Figure 7. $A = S\lambda_1$.

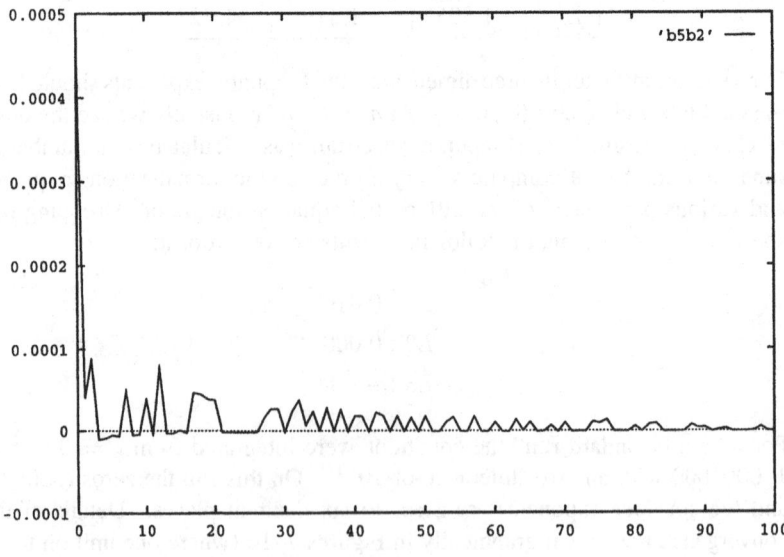

Figure 8. $A = S\lambda_2$.

216

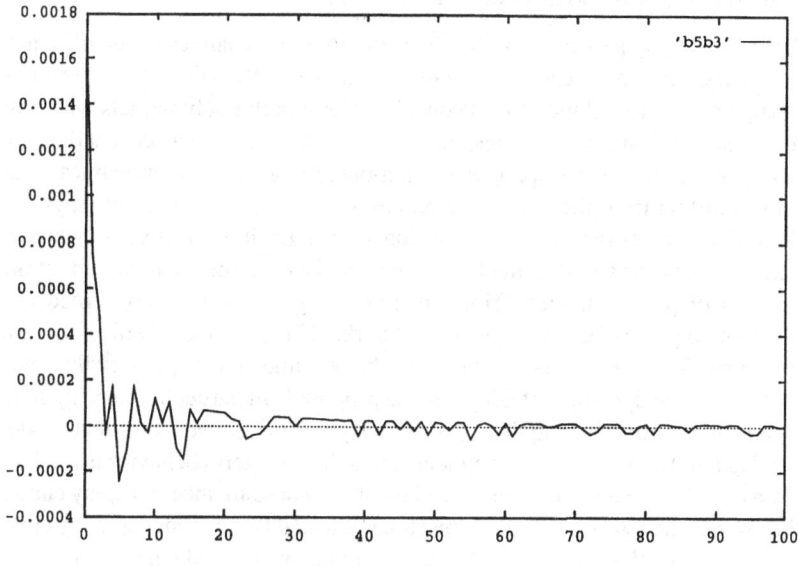

Figure 9. $A = S\lambda_3$.

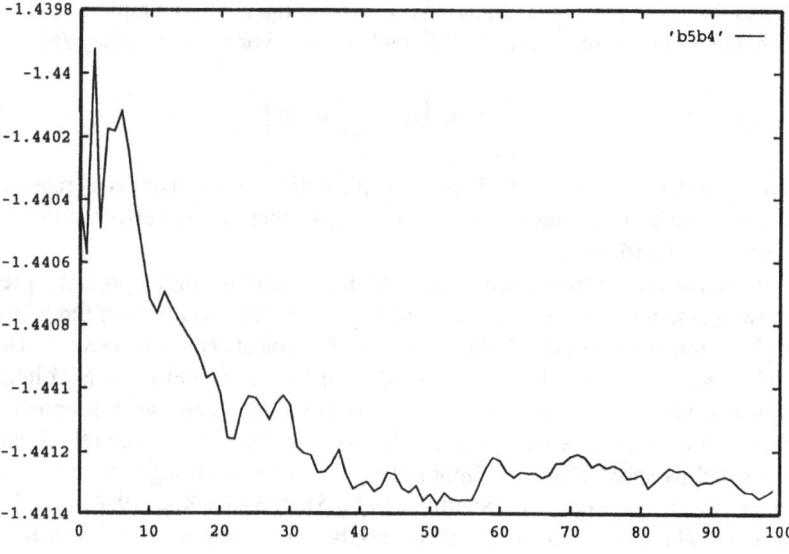

Figure 10. $A = S\lambda_4$.

217

8. RELATION TO LOTKA–VOLTERRA MODELS AND TO OTHER LITERATURE

There are two papers that discuss chaos in different dynamics for equilibration in games: one in an economic context and one in the context of theoretical computer science. Rand (1978) considers Cournot duopoly models where the dynamics is Cournot's best-response map. Where the reaction functions are tent-shaped and have slope greater than one, we get chaotic dynamics. This model differs from the one in the previous section in a number of ways: (1) It is a map rather than a flow that is considered, (2) it is a different dynamics, and (3) when the Cournot model is considered as a game, there are an infinite number of pure strategies. Huberman and Hogg (1988) are concerned with distributed processing on computer networks. The problem of efficient use of resources in a network is modeled as a finite game, and a quasi-evolutionary account of the dynamics of adaptation is proposed. In particular, they argue for chaos in the limit of long delays for a delay differential equation that models the lag in information. The argument is that the long-term behavior is modeled by a difference equation that is in a class all of whose members display chaotic behavior. The setting considered by Huberman and Hogg is conceptually closer to the one in this paper than to that of Rand in that only finite games are considered, but the dynamics is different.

There is a closer connection with ecological models that do not, on the face of them, have much to do with evolutionary-game theory. These are the Lotka–Volterra differential equations, which are intended as simple models of population interactions between different species. For n species, they are:

$$\frac{dx_i}{xt} = x_i \left[r_i + \sum_{j=1}^{n} a_{ij}x_j \right].$$

The x_i are the population densities, the r_i the intrinsic growth or decay rates for a species, and the a_{ij} the interaction coefficients that give the effect of the jth species on the ith species.

The dynamics of two-species Lotka–Volterra systems – either predator–prey or two competing species – is well understood, and the dynamics of three- and higher-dimensional Lotka–Volterra systems is a subject of current research. Unstable cycles are possible in two-dimensional (predator–prey) Lotka–Volterra systems, but chaos is not. In three dimensions, however, several apparent strange attractors have been found. The first was found by Vance (1978) and classified as spiral chaos by Gilpin (1979). "Gilpin's strange attractor" has been extensively studied by Shaffer (1985), Shaffer and Kot (1986), Vandermeer (1991). Other strange attractors have been reported in three-dimensional Lotka–Volterra systems. Arneodo et al. (1980, 1982) use a mixture of numerical evidence and theoretical argument to support the hypothesis of Silnikov-type strange attractors in three and higher dimensions. See also Takeuchi and Adachi (1984) and Gardini et al. (1989). May and Leonard (1975) show that other kinds of wild behavior are possible in Lotka–Volterra systems of three competitors. Smale (1976) shows that for ecological systems modeled by a general class of differential equations (not necessarily Lotka–Volterra), any kind of asymptotic dynamical behavior, including the existence of strange attractors, is possible if there are 5 or more competing species.

There is an intimate connection between the Taylor–Jonker game dynamics and the Lotka–Volterra dynamics, which is established by Hofbauer (1981). A Lotka–Volterra system with n species corresponds to an evolutionary game with $n + 1$ strategies such that the game dynamics on the evolutionary game is topologically orbital-equivalent to the Lotka–Volterra dynamics. To each species in the Lotka–Volterra system, there is a ratio of probabilities of strategies in the game with the same dynamics. Thus it is possible to use known facts about one kind of dynamical system to establish facts about the other. Hofbauer uses the known fact that two-species Lotka–Volterra systems do not admit limit cycles to verify Zeeman's conjecture that 3-strategy evolutionary games do not admit stable limit cycles in the game dynamics. It is thus possible to investigate game-dynamical pathology with an eye toward ecological pathology. The strange attractor of the previous section is, in fact, the game-theoretic counterpart to Gilpin's strange attractor. For a game-dynamical counterpart of the attractor of Arneodo, Coullet, and Tresser we have Example 8:

Example 8

0	−0.6	0	1
1	0	0	−0.5
−1.05	−0.2	0	1.75
0.5	−0.1	0.1	0

A preliminary numerical calculation of the Liapunov spectrum for this system also indicates chaos with one positive Liapunov exponent, although the simulations are not extensive enough to report values with the reliability of those given in Section 7.

9. CONCLUSION

Let us return to the second philosophical thesis with which I began this essay: When the underlying dynamics is taken seriously, it becomes apparent that equilibrium is not the central explanatory concept. Rather, I would take the central dynamical explanatory concept to be that of an *attractor* (or attracting set). Not all dynamical equilibria are attractors. Some are unstable fixed points of the dynamics. In the dynamical system of Example 7 with $a = 5$, there is an unstable equilibrium point that is never seen. And not all attractors are equilibria. There are limit cycles, quasi-periodic attractors, and strange attractors. The latter combine a kind of internal instability with macroscopic asymptotic stability. Thus, they can play the same kind of explanatory role as that of an attracting equilibrium, although what is explained is a different kind of phenomenon.

Even this latter point, however, must be taken with a grain of salt. That is because of the possibility of extremely long transients. In Example 7 with $a = 4$, if we had omitted only the first 50 time units, we would not have eliminated the transient, and the plot would have looked like the strange attractor of Figure 1 rather than one of a limit cycle. If transients are long enough, they may govern the phenomena of interest to us. The concept of an attractor lives at $t = $ infinity, but we do not.

NOTES

1. An earlier version of this paper was delivered at the 1992 meetings of the Philosophy of Science Association and published in Skyrms (1993). The existence of this strange attractor together with a preliminary study of the route to chaos involved was first reported in Skyrms (1992a). This paper contains further experimental results. I would like to thank the National Science Foundation for support, the University of California for a grant of computing time on the Cray computer at the San Diego Supercomputer Center, and Linda Palmer for implementing and running programs to determine the Liapunov spectrum. I would also like to thank Immanuel Bomze, Vincent Crawford, William Harper, and Richard Jeffrey for comments on earlier versions of this paper.
2. But it is not a proper equilibrium. See van Damme (1987) for a definition of proper equilibrium, a proof that if S is an evolutionarily stable strategy, then $\langle S, S \rangle$ is a perfect and proper equilibrium of the associated game, and a great deal of other information about relations between various stability concepts.
3. For flows the sum is replaced with an integral. For three dimensions, there is a spectrum of three Liapunov exponents, each quantifying divergence of the orbit in a different direction.
4. To make the dynamics autonomous, expand the concept of state of the system to include a "memory" of frequencies of past plays.
5. See Bomze (1986), p. 48, or van Damme (1987), p. 222, and note that in the example given, the expected utility of the status quo (= the average population fitness) at a completely mixed equilibrium point must be equal to zero, since for this fitness matrix, the expected utility of strategy 4 is identically zero.
6. For purposes of comparison, the largest Liapunov exponent here is roughly an order of magnitude smaller than that of the Rössler attractor. But the mean orbital period of the attractor is roughly an order of magnitude larger. If we measured time in terms of mean orbital periods, $L1$ would here be of the same order of magnitude $L1$ for the Rössler attractor. Data on the Rössler attractor was obtained from Wolf et al. (1985).

At $a = 6$, although the attractor appears to become geometrically more complex, the Liapunov spectrum is little changed:

$$L1 : 0.009$$
$$L2 : 0.000$$
$$L3 : -1.44$$

REFERENCES

Arneodo, A., P. Coullet, and C. Tresser. 1980. "Occurrence of Strange Attractors in Three Dimensional Volterra Equations." *Physics Letters* 79:259–63.

Arneodo, A., P. Coullet, J. Peyraud, and C. Tresser. 1982. "Strange Attractors in Volterra Equations for Species in Competition." *Journal of Mathematical Biology* 14:153–57.

Bomze, I. M. 1986. "Non-cooperative 2-Person Games in Biology: A Classification." *International Journal of Game Theory* 15:31–59.

Brown, G. W. 1951. "Iterative Solutions of Games by Fictitious Play." In *Activity Analysis of Production and Allocation*. Cowles Commission Monograph. New York: Wiley, pp. 374–76.

Cournot, A. 1897. *Researches into the Mathematical Principles of the Theory of Wealth*. Trans. from the French ed. of 1838. New York: Macmillan.

Crawford, Vincent P. 1989. "Learning and Mixed-Strategy Equilibria in Evolutionary Games." *Journal of Theoretical Biology* 140:537–90. Reprinted in this volume.

van Damme, E. 1987. *Stability and Perfection of Nash Equilibria*. Berlin: Springer.

Eckmann, J. P., and D. Ruelle. 1985. "Ergodic Theory of Chaos and Strange Attractors." *Reviews of Modern Physics* 57:617–56.

Gardini, L., R. Lupini, and M. G. Messia. 1989. "Hopf Bifurcation and Transition to Chaos in Lotka–Volterra Equations." *Mathematical Biology* 27:259–72.

Gilpin, M. E. 1979. "Spiral Chaos in a Predator-Prey Model." *The American Naturalist* 13:306–8.

Guckenheimer, J., and P. Holmes, 1986. *Nonlinear Oscillations, Dynamical Systems and Bifurcations of Vector Fields*. Corrected second printing. Berlin: Springer.

Hirsch, M. W., and S. Smale. 1974. *Differential Equations, Dynamical Systems and Linear Algebra*. New York: Academic Press.

Hofbauer, J. 1981. "On the Occurrence of Limit Cycles in the Volterra-Lotka Equation." *Nonlinear Analysis* 5:1003–7.

Hofbauer, J., and K. Sigmund. 1988. *The Theory of Evolution and Dynamical Systems*. Cambridge: Cambridge University Press.

Huberman, B. A., and T. Hogg. 1988. "The behavior of computational ecologies." In *The Ecology of Computation*, ed. B. A. Huberman, 77–115. Amsterdam: North Holland.

Krishna, V., and T. Sjöström. 1995. "On the Convergence of Fictitious Play." Working paper, Pennsylvania State University.

May, R. M., and W. L. Leonard. 1975. "Nonlinear Aspects of Competition between Three Species." *SIAM Journal of Applied Mathematics* 29:243–53.

Maynard Smith, J. 1982. *Evolution and the Theory of Games*. New York: Cambridge University Press.

Maynard Smith, J., and G. R. Price. 1973. "The Logic of Animal Conflict." *Nature* 146:15–8.

Nachbar, J. H. 1990. " 'Evolutionary' Selection Dynamics in Games: Convergence and Limit Properties." *International Journal of Game Theory* 19:59–89.

Press, J., B. Flannery, S. Teukolsky, and W. Vetterling, 1989. *Numerical Recipes: The Art of Scientific Computing*. Rev. ed. New York: Cambridge University Press.

Rand, D. 1978. "Exotic Phenomena in Games and Duopoly Models." *Journal of Mathematical Economics* 5:173–84.

Rössler, O. 1976. "Different Types of Chaos in Two Simple Differential Equations." *Zeitschrift fur Natuforschung* 31a:1664–70.

Samuelson, L. 1988. "Evolutionary Foundations of Solution Concepts for Finite Two-Player Normal Form Games." In *Proceedings of the Second Conference on Theoretical Aspects of Reasoning about Knowledge*, ed. M. Vardi, pp. 211–26. Los Altos, Calif: Morgan Kaufmann.

Selten, R. 1975. "Reexamination of the Perfectness Concept of Equilibrium in Extensive Games." *International Journal of Game Theory* 4:25–55.

Shaffer, W. M. 1985. "Order and Chaos in Ecological Systems." *Ecology* 66:93–106.

Shaffer, W. M., and M. Kot. 1986. "Differential Systems in Ecology and Epidemiology." In *Chaos: An Introduction*, ed. A. V. Holden. Manchester: University of Manchester Press, pp. 158–78.

Shapley, L. 1964. "Some Topics in Two-Person Games." In *Advances in Game Theory*, Annals of Mathematical Studies, vol. 5, ed. M. Drescher, L. S. Shapley, and L. W. Tucker, pp. 1–28. Princeton: Princeton University Press.

Skyrms, B. 1988. "Deliberational Dynamics and the Foundations of Bayesian Game Theory." In J. E. Tomberlin, ed. *Epistemology*, pp. 345–67. Atascadero, Calif.:Ridgeview.

Skyrms, B. 1989. "Correlated Equilibria and the Dynamics of Rational Deliberation." *Erkenntnis* 31:347–64.

Skyrms, B. 1990. *The Dynamics of Rational Deliberation*. Cambridge, Mass.: Harvard University Press.

Skyrms, B. 1991. "Inductive Deliberation, Admissible Acts and Perfect Equilibrium." In M. Bacharach and S. Hurley, eds., *Foundations of Decision Theory*. Oxford: Blackwell, pp. 220–41.

Skyrms, B. 1992a. "Chaos in Game Dynamics." *Journal of Logic, Language and Information* 1:111–30.

Skyrms, B. 1992b. "Adaptive and Inductive Deliberational Dynamics." In P. Bourgine and B. Walliser, eds., *Economics and Cognitive Science*, pp. 93–107. Oxford: Pergamon Press.

Skyrms, B. 1993. "Chaos and the Explanatory Significance of Equilibrium." *PSA1992*. Vol. 2. Philosophy of Science Association, pp. 374–94.

Smale, S. 1976. "On the Differential Equations of Species in Competition." *Journal of Mathematical Biology* 3:5–7.

Takeuchi, Y., and N. Adachi, 1984. "Influence of Predation on Species Coexistence in Volterra Models." *Mathematical Biosciences* 70:65–90.

Taylor, P., and L. Jonker. 1978. "Evolutionarily Stable Strategies and Game Dynamics." *Mathematical Biosciences* 40:145–56.

Thorlund-Peterson, L. 1990. "Iterative Computation of Cournot Equilibrium." *Games and Economic Behavior* 2:61–75.

Vance, R. R. 1978. "Predation and Resource Partitioning in a One-Predator–Two-Prey Model Community." *American Naturalist* 112:441–48.

Vandermeer, J. 1991. "Contributions to the Global Analysis of 3-D Lotka–Volterra Equations: Dynamic Boundedness and Indirect Interactions in the Case of One Predator and Two Prey." *Journal of Theoretical Biology* 148:545–61.

von Neumann, J., and O. Morgenstern. 1947. *Theory of Games and Economic Behavior*. Princeton: Princeton University Press.

Wolf, A., J. B. Swift, H. L. Swinney, and J. A. Vastano. 1985. "Determining Lyapunov Exponents from a Time Series." *Physica* 16-D:285–317.

Wolfram, S. 1991. *Mathematica* 2d. ed. New York: Addison Wesley.

Zeeman, E. C. 1980. "Population Dynamics from Game Theory." In Z. Niteck and C. Robinson, eds., *Global Theory of Dynamical Systems*, pp. 471–97. Berlin: Springer.